Spectral, Spatial, and
Temporal Properties of Lasers

OPTICAL PHYSICS AND ENGINEERING

Series Editor: William L. Wolfe

Optical Sciences Center
University of Arizona
Tucson, Arizona

1968:
M. A. Bramson
Infrared Radiation: A Handbook for Applications

1969:
Sol Nudelman and S. S. Mitra, Editors
Optical Properties of Solids

1970:
S. S. Mitra and Sol Nudelman, Editors
Far-Infrared Properties of Solids

1971:
Lucien M. Biberman and *Sol Nudelman, Editors*
Photoelectronic Imaging Devices
Volume 1: Physical Processes and Methods of Analysis
Volume 2: Devices and Their Evaluation

1972:
A. M. Ratner
Spectral, Spatial, and Temporal Properties of Lasers

Spectral, Spatial, and Temporal Properties of Lasers

A. M. Ratner
Institute of Low-Temperature Physics and Engineering
Academy of Sciences of the Ukrainian SSR
Khar'kov, USSR

Translated from Russian by
Paul Robeson, Jr.
Chief Translator and Editor
International Physical Index

Supplemented and Edited by
Richard A. Phillips
University of Minnesota
Minneapolis, Minnesota

PLENUM PRESS • NEW YORK – LONDON • 1972

Anatolii Markovich Ratner was born in 1934 in the Ukraine and was graduated from Khar'kov State University in 1956. A candidate of physicomathematical sciences (he defended his dissertation in 1961), he is a theoretical physicist by training, his main field of research being the theory of lasers, predominantly under multimode conditions. He is currently employed at the Institute of Low-Temperature Physics and Engineering, Academy of Sciences of the Ukrainian SSR, Khar'kov, as a senior scientific associate.

The original Russian text, published by Naukova Dumka in Kiev in 1968, has been corrected by the author for the present edition. The translation is published under an agreement with Mezhdunarodnaya Kniga, the Soviet book export agency.

А. М. Ратнер

СПЕКТРАЛЬНЫЕ, ПРОСТРАНСТВЕННЫЕ И ВРЕМЕННЫЕ ХАРАКТЕРИСТИКИ ЛАЗЕРА

SPEKTRAL'NYE, PROSTRANSTVENNYE I VREMENNYE KHARAKTERISTIKI LAZERA

Library of Congress Catalog Card Number 76-167677
ISBN 0-306-30542-9

A Division of Plenum Publishing Corporation
227 West 17th Street, New York, N.Y. 10011

United Kingdom edition published by Plenum Press, London
A Division of Plenum Publishing Company, Ltd.
Davis House (4th Floor), 8 Scrubs Lane, Harlesden, London, NW10 6SE, England

Printed in the United States of America

TRANSLATION EDITOR'S PREFACE

During the decade there were many developments in laser research and numerous applications of the laser were made in fields of science and engineering. Many theoretical and experimental advances were made in the Soviet Union; often they paralleled those taking place in the United States and elsewhere but started from different points, proceeded along different paths, and yielded different insights into the physical processes taking place in the laser.

The present book offers a unified theory of lasers by which the operating characteristics of the laser are described and related to the details of the radiation emitted. Extensive emendations to the original, Soviet edition were supplied by the author and incorporated into the text, and references to the English literature were added to the translation to permit the reader to readily explore topics in greater detail.

Since the Soviet edition's appearance in 1968 one very important area has developed—that of mode locking and picosecond pulse production. This area is so important that no comprehensive work on the laser published in 1972 could neglect it. Accordingly, Chapter XII was added to round off the treatment. I am indebted to my colleague, Professor A. J. Carruthers of the University of Minnesota for many illuminating discussions on mode locking.

Richard A. Philips

FOREWORD TO THE AMERICAN EDITION

The purpose of the present monograph is to examine the physical essence of certain interrelated problems in the theory of solid-state lasers. In striving for the maximum physical clarity of presentation, the author has attempted to use a mathematical device which is adequate for the physics of the problems investigated. This has been expressed in the following way.

In the most general case the electromagnetic field in a resonator is described by a system of two equations for the electric field E and the electric induction D. One of these equations is the Maxwell equation, while the second is the equation for the polarization of the material, written using the density matrix. However, this rather complex mathematical device is inexpedient for use in investigating the fields of a solid-state laser. In fact, the optical bands of condensed media are usually characterized by a width which substantially exceeds the reciprocal of the optical-transition time. In this case it is not difficult to show that the equation for the polarization of the material takes the trivial form $D = \varepsilon E$, where ε is the complex dielectric constant whose imaginary part is proportional to the effective absorption coefficient. Thus, the field is described by the conventional wave equation with complex ε. This wave equation is used in the present monograph to investigate the spatial and spectral distributions of the fields in the resonator. In particular, the radiation spectrum is determined by competition between modes, which is conveniently investigated using the wave equation in the mode-amplitude representation.

The natural resonator oscillations, which are generated simultaneously in fairly large numbers, behave as a unified whole (i.e., competition between them becomes negligible). In this case the wave equation reduces to the elementary kinetic equation for the radiative energy. This kinetic equation is used in the monograph to investigate regular oscillations of multimode radiation. The energy approach is similarly justified in other cases in which the competition between modes is negligible (for example, in the investigation of various threshold phenomena).

Thus, in investigating each problem we use the simplest adequate mathematical device; the author hopes that because of this the monograph has sufficient physical clarity.

In this edition the inaccuracies and typographical errors appearing in the Soviet edition have been corrected to the extent possible. However, it proved to be practically impossible to supplement the book with the results of later research. The presentation of these results is contained in another monograph by the author [108].

A. Ratner

FOREWORD

Many papers on quantum electronics are currently being published, and they are distinguished by a variety of terminology and approach to the problems investigated.

In our monograph we present several closely related problems in the theory of solid-state lasers from a unified point of view. Chapter I presents certain information from luminescence theory; Chap. II presents basic concepts. In Chaps. III-VI we examine the spectral properties and spatial structure of the electromagnetic field in the laser: the third chapter presents linear resonator theory without considering the active medium, while the fourth and fifth chapters develop the nonlinear theory of a practical resonator containing an active medium and analyze the fundamental differences between this case and the linear case, which are expressed, in particular, in the angular and spectral spread of the radiation. Chapter VI examines a resonator with concave mirrors of arbitrary shape from a similar point of view. Chapter VII transforms the wave equation to the "balance-equation" form on the assumption of spatial uniformity of the active medium, and this equation is then used to consider relaxational oscillations of the radiation. The conditions are analyzed under which proper oscillations can actually be observed. Simple expressions are derived for the period, shape, amplitude, and damping of the oscillations. The behavior of the laser in modulated-Q operating modes is closely connected with the characteristics of free oscillations (Chaps. X and XI). The spectral composition of laser radiation is considered in Chap. VIII while taking account of the most essential physical causes of spectral broadening. Chapter IX is devoted to an investigation of threshold phenomena connected with the microinhomogeneity of the active medium.

Quantities having a clear physical meaning are used as the basic variables. A portion of the material presented in the monograph had to be rephrased somewhat in order to retain a unified terminology.

Special attention is devoted to a detailed qualitative description of the physical picture of the phenomena investigated. Brief conclusions are given at the end of each chapter for the convenience of the reader.

This monograph investigates only solid-state lasers with optical pumping; §§ 7, 8, and 18, which are devoted to linear theory and apply equally to gas lasers, are an exception. In view of its limited size, the monograph cannot pretend to exhaustive completeness in the presentation of the theoretical, let alone the experimental, material.

The author is deeply indebted to Professor G. E. Zil'berman and Doctor of Physicomathematical Sciences A. N. Oraevskii for reviewing the manuscript and their valuable comments, as well as to Candidates of Physicomathematical Sciences B. L. Livshits and V. N. Tsikunov for very helpful discussions.

CONTENTS

BASIC NOTATION

a = radius of the generating region of the sample cross section.

a = subscript with which the parameters of the passive shutter are labeled.

c = velocity of light in a vacuum.

$D = \lambda l / a^2$ = small parameter stipulating the diffraction losses.

$\vec{E}$ = electric field vector.

$\mathscr{E}$ = energy of the stimulated emission (in arbitrary units).

$\hbar$ = Planck's constant (divided by 2π).

j = one-half of the number of generated longitudinal modes, § 15.

J = volume energy density of the stimulated emission.

$\vec{k}(k_1, k_2, k_3)$ = wave vector of the light.

$k_\perp = k_1, k_2$ = component of h which is perpendicular to the optic axis.

K = quantum luminescence yield (see (2.3) and (2.4)).

l = resonator length.

l_0 = length of the active rod.

m_3, m_2, m_1 = indices of the longitudinal and transverse modes; $m_\perp = m_1, m_2$.

n = number of excited luminescence centers per unit volume.

n^* = threshold value of n.

$\underline{n}$ and $\bar{n}$ = threshold values of n for an open and closed optical shutter;† $\bar{n}$ coincides with the value of n at the instant generation begins.

$n(\omega)$ = refractive index of the active medium.

N = pump power absorbed per unit volume.

† In the case of a three-level diagram, this value is measured from the value $n_0/2$, where n_0 is the total volume concentration of luminescence centers.

N^* = threshold value of N (in the case of a resonator with modulated Q for a closed optical shutter).
p = ratio between the initial number of excited atoms and n^* (t^*).
$p = \bar{n}/n$ = maximum possible value of p.
$P(\omega) \equiv P_2(\bar{\omega})$ = contour of the luminescence band corresponding to the operating transition, normalized to a unit area.
r = reflectivity of the end mirrors.
R = radius of curvature of a spherical mirror.
s = probability of excitation migration during the time T.
t = time (usually measured from the beginning of generation).
t_0 = characteristic time for the variation of Q.
T = time for a spontaneous optical transition.
t^*, u^* = instant of maximum intensity.
u = the time expressed in $\sqrt{T/v\,\varkappa_1\xi}$ units.
$u(x, y) = z$ = equation for the surface of a concave mirror.
u_{max} = height of a mirror surface or sagitta in the active region of the cross section.
$v = c/n(\omega_0)$ = velocity of light in the material.
$y(t)$ = integral of the effective absorption coefficient along the light path.
Y = oscillation amplitude of the stimulated emission.
Y_1 = initial value of Y.
z = coordinate measured along the optic axis.
γ = additional losses of one of the polarization components.
$\delta = 3/\sqrt{\varkappa_1 k \xi}$ = width of the region in which diffraction is substantial.
δt = width of the intensity peak of the generated light.
$\delta\omega$ = spectral width of the generated radiation.
$\delta\Omega$ = solid angle in which the laser radiation is concentrated.
Δt, Δu = oscillation period of the laser radiation.
$\Delta\omega$ = half-width of the luminescence band.
ε = dielectric constant of the active medium, $\varepsilon_0 = \mathrm{Re}\,\varepsilon$.
ξ = relative amount by which the threshold is exceeded in a four-level diagram $(N - N^*)/N^*$; for a three-level diagram $\xi = [(N - N^*)/N^*] \cdot [(u + \varkappa_1)/\varkappa_1]$.
θ = angular divergence of the laser radiation.
ϑ = angular (admissible) misalignment of the end mirrors.

$\varkappa(\omega, t) =$ effective absorption factor, which allows for light losses and light amplification by the active medium (§ 4).

$\varkappa_1 = \varkappa_0 + (1-r)l =$ losses per unit light path.

$\varkappa_0 =$ absorption coefficient of the basic material.

$\lambda =$ light wavelength.

$\Lambda =$ large parameter of the order of $|\ln \delta\Omega|$.

$\mu(\omega) =$ absorption coefficient in the band corresponding to the working transition for a three-level diagram only.

$\rho(\omega) =$ spectral density of the stimulated radiation.

$\varphi =$ angle between the electric vector and the crystallographic axis.

$\varphi(t) = n^*(t)/n =$ function describing the Q modulation.

$\omega_0 =$ working frequency of the laser.

$\bar{\omega} =$ characteristic frequency interval of the order of the half-width of the luminescence band.

$\omega' = \omega - \omega_0 =$ frequency measured from the working frequency.

Chapter I

CERTAIN INFORMATION FROM LUMINESCENCE THEORY

Nonconducting solids (for example, single crystals) containing a small impurity of luminescent atoms are usually used as the active media for solid-state lasers. Such impurity luminophors can have a relatively narrow luminescence band; this is very essential, since the threshold pump power is proportional to the half-width of the luminescence band. Pure luminescent crystals are impractical active media because of their broad luminescent band (which arises because the transition is between energy bands) and also because of their excessively high absorption. In view of this, we shall restrict our examination to impurity luminescence.

§1. The Absorption and Luminescence Spectra of Impurity Luminophors

Usually the impurity concentration introduced into a nonconducting crystal is so small that the interaction of the impurity atoms with one another can be neglected. Then discrete energy levels, which usually are simply shifted levels of the isolated impurity atom, appear in the interval between the valence band and the conduction band of the dielectric.

The optical transitions between the ground and higher levels usually correspond to rather broad absorption bands (with a width of the order of a thousand cm^{-1}). These impurity absorption bands are shifted toward wavelengths that are long relative to the intrinsic absorption bands of the crystal. Most frequently there are several impurity absorption bands corresponding to transitions to different excited levels.

If the crystal is illuminated by light whose wavelength falls in the impurity absorption band, then the impurity atoms are excited. This can be followed by luminescence (nonradiative deactivation of excited atoms is also possible). Each impurity absorption band is connected with a transition between some pair of levels. A corresponding luminescence band is produced by the reverse transition between the same levels. Sometimes luminescence takes place at wavelengths that are long relative to the corresponding absorption band (the magnitude of this shift is called the Stokes shift).

Because of nonradiative transitions between excited levels the illumination of the crystal in different absorption bands can be accompanied by luminescence in the same band, which corresponds to a transition from the lowest of the excited levels. With increasing temperature, the probability of nonradiative transitions increases and the luminescence spectrum ceases to depend on the impurity absorption band in which the crystal is excited. To illustrate the above Fig. 1 shows the absorption and luminescence spectra of certain thallium-activated alkali-halide crystals at room temperature; the crystal luminesces in the band shown in the figure, regardless of the absorption band in which it is illuminated [83, 89].

The broadening of the energy levels in the band is caused by the interaction of an impurity center with vibrations of the crystal lattice. In the language of quantum mechanics this means that a

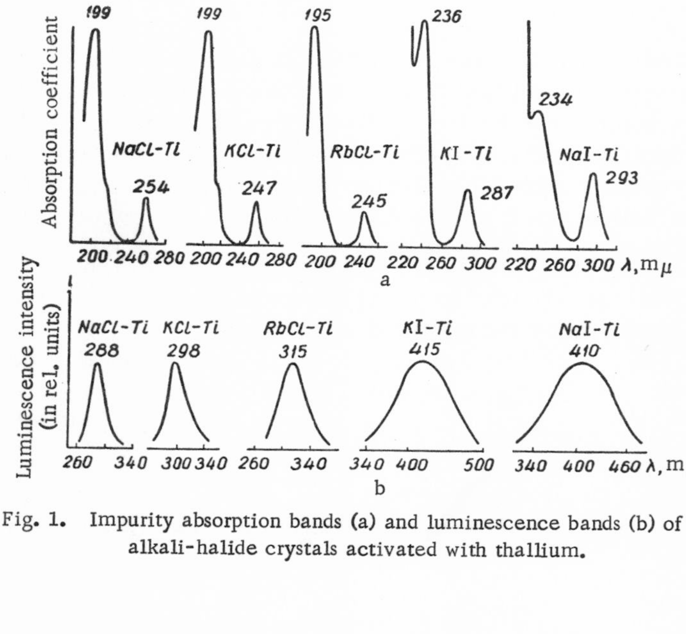

Fig. 1. Impurity absorption bands (a) and luminescence bands (b) of alkali-halide crystals activated with thallium.

transition between electronic levels of the impurity atoms is accompanied by the emission of a certain number of phonons [117, 118]. The phonon involved can vary within fairly wide limits. However, if the problem is examined classically (we shall do precisely this for simplicity in reasoning), then the existence of the luminescence and absorption bands is connected with the fact that the positions of the levels change continuously due to the motion of the lattice ions during the process of light emission or absorption.

The absorption coefficient, which is connected with transitions between the impurity levels 1 and 2, is conveniently represented in the following form [56, 57]:

$$\mu(\omega) = \frac{4\pi^2 n_1 |\vec{d}_{12}|^2 \omega}{3\hbar c n(\omega)} P_1(\omega) \tag{1.1}$$

Here n_1 is the number of absorption centers per unit volume which are located at the level 1†; ω is the light frequency; $\vec{d}_{12}$ is the matrix element of the electric dipole moment‡ between states 1 and 2; c is the velocity of light in vacuum; $n(\omega)$ is the refractive index of the material. Here $P_1(\omega)$ designates the function describing the shape of the absorption band; this function is normalized per unit area:

$$\int_{-\infty}^{\infty} P_1(\omega)\, d\omega = 1. \tag{1.2}$$

Compared with $P_1(\omega)$, the remaining factors in Eq. (1.1) depend weakly on frequency, and this dependence can be negected. The quantity $P_1(\omega)$ is determined by the motion of the lattice ions, and the remaining factors basically describe the motion of the electrons and thus do not reveal a noticeable temperature dependence. The function $P_1(\omega)$ is bell-shaped, and its half-width $\Delta\omega_1$ increases in proportion to $\sqrt{T}$ with increasing temperature T.

Analogously, the spectral intensity of the luminescence per unit volume can be represented in the form

$$J(\omega) = \frac{4}{3} \cdot \frac{n_2 |\vec{d}_{21}|^2 n(\omega) \omega^4}{c^3} P_2(\omega), \tag{1.3}$$

† Usually this is the ground-state level.

‡ If the electric dipole transition $1 \to 2$ is forbidden, then $\vec{d}_{12}$ should be understood to represent the magnetic dipole or electric quadrupole moment.

where n_2 is the number of excited impurity atoms per unit volume. The function $P_2(\omega)$ describes the shape of the luminescence band and is normalized to a unit area:

$$\int_{-\infty}^{\infty} P_2(\omega)\, d\omega = 1. \tag{1.4}$$

The functions $P_1(\omega)$ and $P_2(\omega)$ usually do not coincide.

In order to carry out a qualitative investigation of the form of the functions P_1 and P_2, their interrelationship and their temperature dependence, let us examine a very simplified model of an impurity atom in a crystal: a diatomic molecule in thermodynamic equilibrium with a thermal reservoir. In this model the position of the electronic levels depends solely on one parameter – the distance R between nuclei. On the other hand the position of the impurity-atom levels in the crystal is determined by the coordinates of many lattice ions. In spite of its simplicity, this model provides a good representation of the basic qualitative relationships obeyed by the absorption and the luminescence spectrum. This is explained by the fact that the optical characteristics of an impurity atom in the crystal are actually determined by the interaction of the impurity atom with a small number of nearest neighbors [57, 69].

For simplicity we shall examine the motion of the nuclei classically and shall use parabolas to depict the electronic states of the molecule (Fig. 2). We designate the equilibrium positions of the nuclei in states 1 and 2 by R_1 and R_2 and write the dependence of the

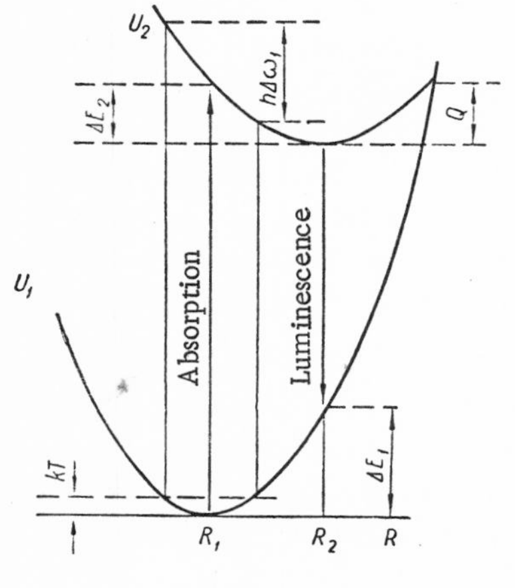

Fig. 2. Electronic states for the model of a diatomic molecule existing as an impurity in a host lattice.

position of the terms on the distance between nuclei $U_1 = K_1(R - R_1)^2/2$ (the ground state) and $U_2 = E_0 + K_2(R - R_2)^2/2$ (the term of the excited state). The curvature of the parabolas and the equilibrium position of the nuclei depend, in general, on the electronic state.

Since the frequency of revolution of the electrons in their orbits (i.e., the electron binding energy divided by $\hbar$) considerably exceeds the frequency of the lattice vibrations, it can be assumed that the electronic transitions are accomplished for fixed positions of the nuclei (the Frank-Condon principle). An electronic transition is preceded by thermodynamic equilibrium; therefore, the transition $1 \to 2$ occurs for values of R close to R_1, while the reverse transition $2 \to 1$ occurs for R close to R_2. From Fig. 2 it is evident that the energy of a quantum absorbed for $R = R_1$ exceeds the energy emitted for $R = R_2$ by the amount $(K_1 + K_2)(R_1 - R_2)^2/2$. From this we find the Stokes shift, i.e., the distance on the frequency scale between the absorption and luminescence maxima:

$$\Delta\omega_{St} = \frac{1}{2\hbar}(K_1 + K_2)(R_1 - R_2)^2. \qquad (1.5)$$

Obviously, the Stokes shift is always positive (this is similarly clear from the second law of thermodynamics).

The difference between the energies of an absorbed quantum and an emitted quantum is converted into heat. This occurs as follows. Immediately after the transition $1 \to 2$ the molecule acquires a vibrational energy ΔE_2 (Fig. 2), which considerably exceeds kT (k is Boltzmann's constant, T is temperature). Thermodynamic equilibrium is established in a few vibration periods of the nuclei (i.e., during 10^{-12} to 10^{-13} sec), and the excess vibrational energy is redistributed between neighboring nuclei (which are represented by the thermal reservoir in our model). It should be noted that thermodynamic equilibrium is always established not only in the ground state but also in an excited electronic state whose lifetime is several orders of magnitude greater than the vibration period of the nuclei.

The transition $1 \to 2$ occurs with overwhelming probability for values of R which are determined from the condition $U_1 \equiv K_1(R - R_1)^2/2 \sim kT$ (Fig. 2). From this we have $|R - R_1| \sim \sqrt{2kT/K_1}$, and, as is evident from the figure, the width of the absorption band is

$$\Delta\omega_1 \sim \frac{1}{\hbar}\cdot\frac{K_2}{\sqrt{K_1}}\,|R_1 - R_2|\sqrt{kT}. \tag{1.6}$$

Analogously, we have the following results for the width of the luminescence band:

$$\Delta\omega_2 \sim \frac{1}{\hbar}\cdot\frac{K_1}{\sqrt{K_2}}\,|R_1 - R_2|\sqrt{kT}. \tag{1.7}$$

For the particular case in which $K_1 = K_2$ (i.e., the elastic constant does not change during an electronic transition) it follows from these formulas that $\Delta\omega_1 = \Delta\omega_2$. However, it can be shown that not only the widths but also the shapes of the absorption and luminescence bands coincide. More precisely, in the case of conserved elastic constants, the functions P_1 and P_2 describing the shape of the optical bands are related by the equation [55] (Fig. 3)

$$P_1(\omega_e + \omega) = P_2(\omega_e - \omega), \tag{1.8}$$

where $\hbar\omega_e = E_0$ is the difference between the minimum positions of the terms 2 and 1 (the quantity ω_e is called the frequency of a purely electronic transition). Equation (1.8) is called the mirror-symmetry law (this law was discovered experimentally by Levshin [39]).

The mirror-symmetry law is satisfied only in those rather rare cases in which the electronic transition is not accompanied by a change in the elastic constant. In general the luminescence band is either narrower than the absorption band (for $K_1 < K_2$) or wider than it (for $K_1 > K_2$); it is more frequently wider, since the elastic constants are usually greater in the ground state than in the excited state.

From Eqs. (1.5) – (1.7) we obtain the relationship between the half-width of the optical impurity bands $\Delta\omega$ and the magnitude

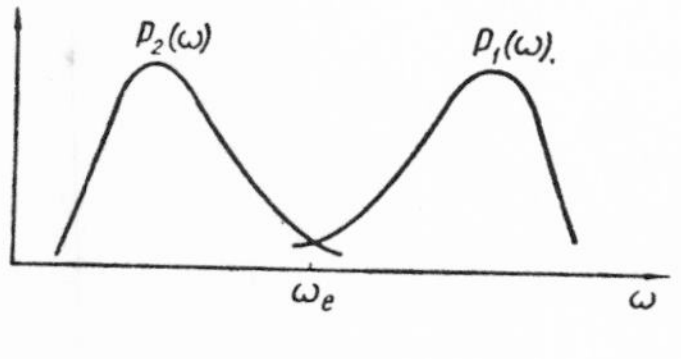

Fig. 3. "Mirror symmetry" of the absorption and luminescence bands.

of the Stokes shift:

$$kT\Delta\omega_{St} \sim \hbar\Delta\omega^2. \tag{1.9}$$

Both the Stokes shift and the finite spectral width of the optical impurity bands are connected with the fact that an electronic transition in an impurity atom is accompanied by a change in the equilibrium position of the nuclei, and this is in turn caused by a change in the forces of interaction between the impurity atom and nearest-neighbor atoms. Usually, the electrons of the outer shell of the impurity atom participate in optical transitions, so that the interaction forces depend essentially on the electronic state. This leads to a large magnitude of the Stokes shift and a large half-width of the optical bands. The large spectral width makes the overwhelming majority of impurity luminophors impractical for use in lasers. In order to obtain appropriate active media, impurity atoms are used for which the optical electrons belong to the inner shell (for example, the rare-earth elements). The state of the electrons of the inner shells does not noticeably affect the forces of interaction between an impurity atom and the lattice ions, and therefore the optical bands turn out to be fairly narrow [121]. For example, for glass with a neodymium impurity the working transition corresponds to a luminescence band with a half-width of approximately 30 cm^{-1}, which is about one order of magnitude smaller than the width of the bands shown in Fig. 1.

It can be shown that for the case in which the mirror-symmetry law is satisfied the functions $P_1(\omega)$ and $P_2(\omega)$ become Gaussian functions at a sufficiently high temperature:

$$P_2(\omega) = \frac{2\sqrt{\ln 2}}{\sqrt{\pi}\Delta\omega_2} e^{-\frac{4\ln 2}{\Delta\omega_2^2}(\omega-\omega_0)^2} = \frac{0.94}{\Delta\omega_2} e^{-\frac{2.77}{\Delta\omega_2^2}(\omega-\omega_0)^2} \tag{1.10}$$

(P_1 has an analogous form). Here ω_0 is the frequency corresponding to the maximum of the luminescence band; $\Delta\omega_2$ is its half-width.

In the general case the impurity optical bands deviate somewhat from the Gaussian shape [34, 68, 69], but this difference is not very sharp, as is evident from Fig. 1.

The broadening of the luminescence line into a band was examined above; this broadening is connected with a change in the equilibrium position of the nuclei which accompanies an electronic transition, and it has a classical interpretation and is conserved for $h \to 0$. However, if the optical electrons belong to inner shells and the electronic transition is not accompanied by a change in the equilibrium of the nuclei and of the elastic constants, then the classical broadening examined above vanishes, as is evident from Eqs. (1.6) and (1.7). In this case it is necessary to consider the small broadening of the spectral lines which is connected with quantum effects – namely, with the interaction between optical electrons and phonons which lead to a finite lifetime of the optical electrons at electron-phonon levels. This lifetime is comparable with $1/\omega_l$ (ω_l is the vibration frequency of the nuclei). On this basis, according to the well-known uncertainty principle, the width of the spectral line must be of the order of $\hbar\omega_l$ and considerably less than the width of ordinary impurity optical bands, which is connected with the change in the equilibrium position of the nuclei during an electronic transition. As is well known, optical lines which have been broadened due to the finite lifetime of the electrons have a Lorentz shape. Thus, the optical impurity bands corresponding to transitions in the inner shells, which are fairly well screened, have a small width and a shape close to the Lorentz shape.† The conventional impurity bands which correspond to electronic transitions in the outer shells have a large width and a shape which is more or less close to Gaussian.

Although in the general case the shape of the optical impurity bands differs from the Gaussian, we shall nevertheless assume that they are Gaussian and make use of Eq. (1.10) hereafter for the sake of brevity. This simplification is justified, since we shall be interested only in the vicinity of the luminescence-band maximum, where the shape of the band can be approximated by a Gaussian function rather than in the entire band. Consideration of the true shape of the band would lead merely to a small change in the numerical coefficient in the exponents of Eq. (1.10).

From Eqs. (1.6) and (1.7) it follows that the width of the optical band is proportional to the square root of the temperature. This conclusion is valid at a fairly high temperature (for practical purposes at room temperature and above) at which the motion of the nuclei can be treated classically. For $T \to 0$ the classical ex-

† Experimental investigation of the R-lines corresponding to the absorption of ruby, which was carried out by Schawlow [122, 123] and by Bel'skii and Mukhamedova [13]. The latter showed that the shape of these lines is close to the Lorentz shape.

amination becomes inapplicable, and the width of the bands remains finite. If we examine the motion of nuclei from the quantum-mechanical standpoint, then the spreading of the absorption or luminescence line into a band is connected with the fact that the electronic transition is accompanied by transitions between various vibrational levels. Although at absolute zero the crystal is at the lowest vibrational level, transitions are possible from this level to various vibrational levels of another electronic state (in the stipulated case the selection rules for the transition of a simple harmonic oscillator lose their validity). This is what leads to a nonzero width of the optical band at absolute zero.

Quantum effects can be considered very simply with an accuracy which is satisfactory for practical purposes: the temperature T in all of the equations is replaced by the "effective temperature" $T^* = (\hbar\omega_l/2)\coth(\hbar\omega_l/2kT)$, where ω_l is the characteristic frequency of the lattice vibrations and is somewhat lower than the Debye frequency. For $T \to \infty$ the effective temperature coincides with the true temperature, and it remains finite for $T \to 0$ (Fig. 4). A reduction of the temperature below the value $\hbar\omega_l/4k$ has no practical effect on the optical characteristics of the impurity atoms.

In conclusion we examine the absorption of light by a crystal in which a portion of the impurity atoms is in the excited electronic state 2. For this purpose we note that Eq. (1.1) can be derived by examining and summing all elementary transitions between the vibrational levels of the ground state 1 and the excited electronic state 2. Assume now that a certain number of atoms n_2 is in the state 2; then stimulated optical transitions $2 \to 1$ become possible. These produce amplification (or negative absorption) of light that is trans-

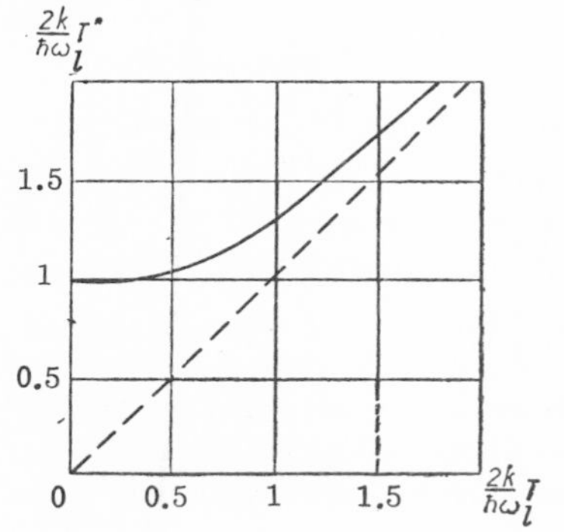

Fig. 4. Dependence of the effective temperature T* on the temperature T.

mitted through the crystal. According to the well-known Einstein rule, the probability of an elementary stimulated transition between vibrational levels of the states 2 and 1 is equal to the probability of an elementary stimulated transition in the opposite direction, namely absorption. Therefore, the summation of all elementary transitions 2 → 1 produces a result analogous to Eq. (1.1). The absorption coefficient turns out to be equal to

$$\mu'(\omega) = -\frac{4\pi^2 n_2 |\vec{d}_{12}|^2 \omega}{3\hbar c n(\omega)} P_2(\omega). \tag{1.11}$$

Equation (1.11) differs from (1.1) by the factor $P_2(\omega)$, which describes the shape of the luminescence band. This is related to the fact that in summing the transitions between vibrational levels thermodynamic averaging is carried out over the vibrational levels of the original electronic state (the state 2 in the case given). The impurity atom is in the electronic state 2 before both spontaneous and stimulated transitions of the 2 → 1 type; regardless of whether these transitions are stimulated or spontaneous, they occur in the same frequency band with a maximum at the point $\omega = \omega_0$ corresponding to the equilibrium position of the nuclei (i.e., to their most probable configuration) in the state 2.

Adding (1.1) and (1.11), we find the coefficient of impurity absorption of unpolarized light by a partially excited impurity luminophor:

$$\tilde{\mu}(\omega) = \frac{4\pi^2 |\vec{d}_{12}|^2 \omega}{3\hbar c n(\omega)} [n_1 P_1(\omega) - n_2 P_2(\omega)]. \tag{1.12}$$

This equation will be used hereafter.

§2. Nonradiative Transitions between Electronic Levels of an Impurity Atom

In the previous section we examined optical bands of impurity atoms in the adiabatic approximation making the assumption that the motion of the nuclei was infinitely slow compared to the motion of the electrons. The adiabatic approximation permits the motion of the electrons and electronic transitions for fixed coordinates of the nuclei. However, in view of the nonzero velocity of the nuclei, the adiabatic approximation cannot be completely exact; therefore, the electronic states examined in the adiabatic approximation are not rigorously stationary. When nuclear motion is taken into account nonradiative transitions are possible between the electronic states of an impurity atom [33, 37, 120].

In order to carry out a qualitative investigation of the probability of nonradiative transitions as a function of temperature and the mutual arrangement of the levels, we turn once more to Fig. 2. From the energy conservation law it follows that a nonradiative transition is possible only for those nuclear configurations for which the distance between levels decreases to a value of the order of $\hbar\omega_l$ (i.e., the levels practically intersect). From Fig. 2 it is obvious that the probability of nuclear configurations for which the levels intersect is proportional to exp $[-Q/kT]$, where Q is a constant which depends on the mutual arrangement of the levels and is called the activation energy of a nonradiative transition. Thus, in the classical case (the temperature exceeds the Debye temperature of the crystal) the probability of a nonradiative transition between levels has the form

$$w = se^{-\frac{Q}{kT}}. \tag{2.1}$$

This quantity is referred to a unit time. The constant s has the dimensionality sec^{-1}, and it is sometimes erroneously identified with the Debye frequency of lattice vibrations; actually, it can differ from the Debye frequency by several orders of magnitude on the small side [47].

Equation (2.1) is applicable both to transitions from the higher level 2 to the lower level 1, and to the reverse transitions $1 \rightarrow 2$. The probabilities of these transitions are connected with the exact relationship that is derived from the detailed-equilibrium principle

$$w_{12} = w_{21} \exp\left[-\frac{E_0}{kT}\right], \tag{2.2}$$

where E_0 is the difference between the minimum values of the terms 2 and 1. Usually, the distance between the electronic terms considerably exceeds kT, so that transitions from the lower level to the higher level have a vanishingly small probability. For this reason it is usually the practice to examine only those nonradiative transitions which occur from the upper level to the lower level.

Equation (2.1) can describe either a transition between the levels of excited states or a transition from an excited state to the ground-state level. In the first case nonradiative transitions

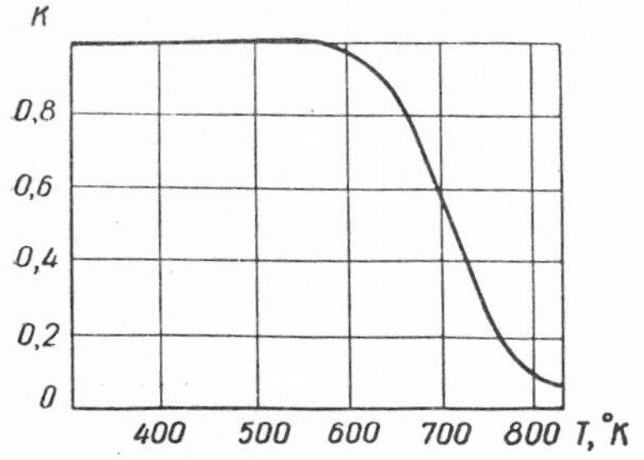

Fig. 5. Experimental dependence of the quantum luminescence yield on temperature for a KCl crystal activated with thallium.

lead to predominant deexcitation of excited atoms from the lower levels of the excited state, while in the second case they lead to a reduction of the quantum luminescence yield K. The quantum yield can be represented in the form

$$K = \frac{w_{\mathrm{rad}}}{w + w_{\mathrm{rad}}} \tag{2.3}$$

(w_{rad} is the probability of a spontaneous optical 2 → 1 transition per unit time), or, since w_{rad} is practically independent of temperature,

$$K = \frac{1}{1 + \mathrm{const} \cdot \exp\left[-\frac{Q}{kT}\right]}. \tag{2.4}$$

Figure 5 shows the characteristic curve for the temperature dependence of the quantum luminescence yield [77].

The classical equations (2.1) and (2.4) are applicable only at temperatures above the Debye temperature,† but they are qualitatively valid at low temperatures, provided only that T is replaced by the "effective temperature" T* (see §1). At absolute zero the value of T* (and therefore the value of w also) remains nonzero; physically, this is connected with zero vibrations of the crystal lattice [33, 57].

It remains for us to examine the dependence of the activation energy on the shape and mutual arrangement of the electronic levels. Without carrying out the calculations, we restrict outselves to establishing the following qualitative relationships [58], which can be illustrated using the diatomic-molecule model. If the distance E_0

† Equation (2.2) is exact and is applicable at any temperature.

between levels is increased, then the activation energy Q increases approximately in proportion to E_0 (Figs. 6a and 6b). As a result of this, partial or complete thermal equilibrium is usually established between close levels. Then, if the transition of an impurity atom from the upper state to the lower state is accompanied by an increase in the elastic constants which correspond to its shift relative to its nearest neighbors, then Q decreases noticeably in comparison with the case of "mirror symmetry" in which the elastic constants are conserved (Fig. 6c). Finally, the shift of the equilibrium position of the nuclei during an electronic transition similarly facilitates the reduction of Q, although not so strongly (Fig. 6d).

Thus, the activation energy of nonradiative transitions is determined basically by the same parameters that determine the shape of the optical bands. However, the role played by these parameters turns out to be somewhat different. In fact, even though the width of the optical impurity bands $\Delta\omega$ and the probability of nonradiative transitions are determined by the interaction of the impurity atom with the lattice vibrations, it is nevertheless true that $\Delta\omega$ depends predominantly on the shift of the equilibrium position, while w depends on the change of the other constants during an electronic transition and on the distance between levels. This difference is explained by the fact that the shape of the optical bands is determined by the motion of the nuclei in a small vicinity of the equilibrium position, while the probability of nonradiative transitions is determined by their motion in the region far from the equilibrium position in which the levels intersect.

§3. Nonradiative Energy Transfer between Impurity Atoms

We have examined nonradiative transitions between levels of the same impurity atom. If the distance between impurity atoms is

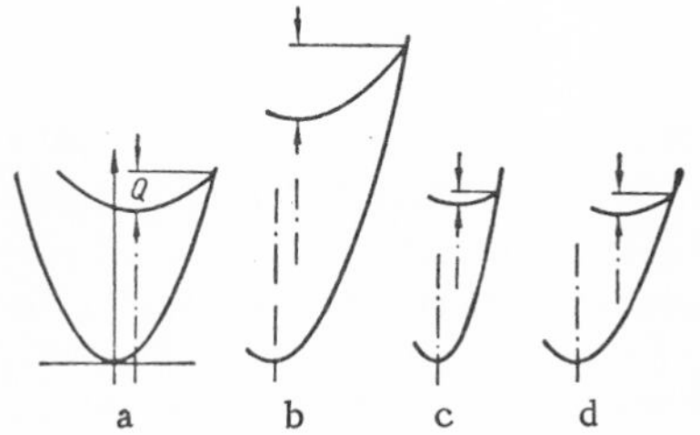

Fig. 6. Diagrams illustrating the qualitative dependence of the activation energy of nonradiative transitions between electronic impurity terms on the shape and mutual arrangement of the levels.

not too great, then under certain conditions nonradiative transitions connected with the transfer of excitation energy from one impurity atom to another are similarly possible. The phenomenon of sensitized luminescence, which is excited by illuminating the crystal in the absorption band of the sensitizer, is based on nonradiative energy transfer from one impurity (sensitizer) to another (activator); on the other hand, deexcitation occurs in the luminescence band of the activator.

Nonradiative energy transfer from a sensitizer atom (S) to an activator atom (A) has been examined in the papers by Förster [84] and Dexter [82]. We adopt a more compact presentation of this theory by using the functions $P_1(\omega)$ and $P_2(\omega)$ which were introduced in § 1 and describe the shape of the optical bands.†

First let us note that the function $P_2(\omega)$, which describes the shape of the luminescence band, is proportional to the probability of those nuclear configurations for which the distance between levels 2 and 1 is equal to $\hbar\omega$, i.e.,

$$P_2(\omega) = \hbar \int \mathfrak{P}_2(x)\,\delta(U_2(x) - U_1(x) - \hbar\omega)\,dx, \qquad (3.1)$$

where x is the aggregate of nuclear coordinates; $\mathfrak{P}_2(x)$ is the x-distribution function for a lattice which is in thermodynamic equilibrium with the excited atom (the function $\mathfrak{P}_2$ is normalized to an integral equal to unity). It is not difficult to verify the proposition that the normalization of the function (3.1) satisfies the condition (1.4).

The form of the distribution function of the nuclear coordinates for a lattice which is in thermodynamic equilibrium with an impurity atom depends on the electronic state this atom is in. Assume $\mathfrak{P}_1(x)$ is the distribution function for a lattice containing an impurity atom in the ground state. Then the function P_1 which stipulates the shape of the absorption band is written in the form

$$P_1(\omega) = \hbar \int \mathfrak{P}_1(x)\,\delta(U_2(x) - U_1(x) - \hbar\omega)\,dx. \qquad (3.2)$$

Assume the crystal contains two impurity atoms S (sensitizer) and A (activator). We introduce the eigenfunction of the electrons of

† Although the treatment is carried out for the classical case, the results are applicable at any temperatures. In particular, the treatment is applicable to the transfer of excitation energy between atoms of the same impurity.

the sensitizer atom in the ground state ($\psi_1(\rho)$) and in the excited state ($\psi_2(\rho)$), as well as the corresponding energy eigenvalues ε_1 and ε_2. We write the eigenfunctions and energy levels of the electrons of the activator atom $\Psi_1(r)$, E_1 (for the ground state) and $\Psi_2(r)$, E_2 (for the excited state) analogously. The eigenfunctions and energy levels depend on the coordinates of the nuclei.

The energy transfer S → A is accomplished due to the effect of a perturbation connected with Coulomb interactions of the electrons of the sensitizer and activator atoms. The probability of transitions due to the effect of this perturbation per unit time has the form

$$e^2 \frac{2\pi}{\hbar}\left|\int \psi_2^*(\vec{\varrho})\,\Psi_1^*(\vec{r})\,\psi_1(\vec{\varrho})\,\Psi_2(\vec{r})\,\frac{\delta(\varepsilon_2+E_1-\varepsilon_1-E_2)}{|\vec{r}-\vec{\varrho}|}\,d\vec{r}\,d\vec{\varrho}\right|^2. \quad (3.3)$$

For simplicity in notation we choose one electron each on the S and A atoms and neglect the exchange interaction, which is exponentially small for $R \gg a$ (a is the lattice constant; R is the distance between the S and A atoms).

In Eq. (3.3) we expand the factor $|\vec{r}-\vec{\rho}|^{-1}$ in powers of the ratios $(\vec{r}-\vec{r}_0)/R$ and $(\vec{\rho}-\vec{\rho}_0)/R$, where $\vec{r}_0$ and $\vec{\rho}_0$ are the coordinates of the nuclei of the A and S atoms. We use $R = |\vec{r}_0 - \vec{\rho}_0|$ to designate the distance between nuclei. Since the functions $\psi_1(\vec{\rho})$ and $\psi_2(\vec{\rho})$ are orthogonal with respect to one another just as the functions $\psi_1(\vec{r})$ and $\psi_2(\vec{r})$ are, it follows that the greatest contribution to the integral is made by the quadratic terms in the expansion, which contain the products of the components of the vectors $\vec{r}-\vec{r}_0$ and $\vec{\rho}-\vec{\rho}_0$. The integrals of these terms are expressed in terms of products of the matrix elements of the dipole moment:

$$\vec{d}_{12}^{A} = e\int \Psi_1^*(\vec{r})\,\vec{r}\,\Psi_2(\vec{r})\,d\vec{r}, \quad \vec{d}_{12}^{S} = e\int \psi_1^*(\vec{\varrho})\,\vec{\varrho}\,\psi_2(\vec{\varrho})\,d\vec{\varrho}.$$

After averaging over the directions $\vec{d}_{12}^{A}$ and $\vec{d}_{12}^{S}$, the quantity (3.3) takes the form

$$w_{SA} = \frac{4\pi}{3\hbar}\cdot\frac{|\vec{d}_{12}^{A}|^2\,|\vec{d}_{12}^{S}|^2}{R^6}\,\delta(\varepsilon_2-\varepsilon_1-E_2+E_1). \quad (3.4)$$

In order to find the average probability of energy transfer per unit time, Eq. (3.4) must be averaged over the configurations of the nuclei. Under these conditions we will assume that the probability of energy transfer per unit time is considerably less than the Debye frequency of the lattice, so that thermodynamic equilibrium is established between the lattice and the impurity atom.

The quantity x_S designates the aggregate of coordinates of the ions surrounding the atom S, while x_A represents the same aggregate for the atom A. The quantities ε_1 and ε_2 depend on x_S, while E_1 and E_2 depend on x_A. For a sufficiently large R it is possible to assume that x_S and x_A are statistically independent. Assume $\mathfrak{P}_2(x_S)$ is the coordinate distribution function for the ions surrounding the excited atom S, while $\mathfrak{P}_1(x_A)$ represents the same function for the atom A, which is in an excited state. Then we have the following equation for the mean probability of excitation energy transfer S → A per unit time:

$$\bar{w}_{SA} = \iint w_{SA}\mathfrak{P}_2(x_S)\,\mathfrak{P}_1(x_A)\,dx_S\,dx_A = \frac{4\pi}{3\hbar R^6}\,|\vec{d}^{A}_{12}|^2|\vec{d}^{S}_{12}|^2\int\delta(\varepsilon_2 - \varepsilon_1 - E_2 + E_1)\,\mathfrak{P}_2(x_S)\,\mathfrak{P}_1(x_A)\,dx_S dx_A = \frac{4\pi}{3R^6}\,|\vec{d}^{A}_{12}|^2\,|\vec{d}^{S}_{12}|^2 \times$$

$$\times\int_{-\infty}^{\infty} d\omega \int \mathfrak{P}_1(x_A)\,\delta(E_2 - E_1 - \hbar\omega)\,dx_A \int \mathfrak{P}_2(x_S)\,\delta(\varepsilon_2 - \varepsilon_1 - \hbar\omega)\,dx_S. \quad (3.5)$$

Taking the definitions (3.1) and (3.2) into account, we find†

$$\bar{w}_{SA} = \frac{4\pi}{3\hbar^2 R^6[n(\omega)]^4}\,|\vec{d}^{S}_{12}|^2\,|\vec{d}^{A}_{12}|^2\int P_1^A(\omega)\,P_2^S(\omega)\,d\omega. \quad (3.6)$$

Thus, the probability of S → A energy transfer is proportional to the integral of the overlap between the absorption spectrum of the activator and the luminescence spectrum of the sensitizer. The physical meaning of this result is quite obvious. Nonradiative S → A energy transfer is possible only for those nuclear configurations for which the energy $\varepsilon = \varepsilon_2 - \varepsilon_1$ released during a transition of an S atom from the excited state to the ground state coincides with the energy $E = E_2 - E_1$ required to excite an A atom. The probability that an excited S atom will have the stipulated energy ε for a transition to the ground state is proportional to the function $P_2^S(\varepsilon/\hbar)$; analogously, the probability that an A atom will have the

† In [82, 84] a refractive index is introduced in order to take account of the polarization of the crystal by the electrons of the impurity atom.

stipulated energy required for a $1 \rightarrow 2$ transition is proportional to $P_1(E/\hbar)$. The probability that the quantities $\varepsilon = \varepsilon_2 - \varepsilon_1$ and $E = E_2 - E_1$ will coincide is proportional to the integral of the overlap between the functions $P_2^S(\omega)$ and $P_1^A(\omega)$.

Taking account of the Stokes shift for the optical bands of both the activator and sensitizer, we arrive at the conclusion that the $S \rightarrow A$ energy transfer has a considerable probability, provided that the distance between the upper and lower states is somewhat greater for the sensitizer atom than for the activator atom (this difference must coincide approximately with the half-sum of the Stokes shift for the S and A atoms, multiplied by $\hbar$). The reverse energy transfer $A \rightarrow S$ has a vanishingly small probability for such an arrangement of the levels, since the activator luminescence band has practically no overlap with the sensitizer absorption band (this is also clear from the second law of thermodynamics). Therefore, the excitation is localized on the activator atoms, and the crystal is luminescent in the luminescence band of the activator when it is excited in the absorption band of the sensitizer (sensitized luminescence).

Equation (3.6) can also describe the transfer of excitation energy between identical impurity atoms. In this case the probability of energy transfer obviously becomes more considerable as the Stokes shift decreases.

It can happen that for the transfer investigated one of the matrix elements $\vec{d}_{12}^S$ and $\vec{d}_{12}^A$ is zero. If, for example, $\vec{d}_{12}^A = 0$ for $\vec{d}_{12}^S \neq 0$, then the expansion of the quantity $|\vec{r} - \vec{\rho}|^{-1}$ must retain terms which are quadratic in $\vec{r} - \vec{r}_0$. In this case the probability of energy transfer is, as previously, proportional to the integral over the overlap of the spectrum; however, it is less in the ratio $(a/R)^2 \sim c_A^{2/3}$ than it is in the case of an allowed dipole transition (a is the lattice constant; c_A is the activator concentration). If the dipole transition is simultaneously forbidden in both the A and S atoms (but quadrupole transitions are allowed), then the probability of energy transfer decreases in the ratio $(a/R)^4$, etc.

Note that a forbidden dipole transition (but one that is an allowed quadrupole transition) leads to a reduction of the probability of deexcitation of an excited atom in the ratio $(a/\lambda)^2 \sim 10^{-6}$ (λ is the light wavelength); the probability of energy transfer decreases only in the ratio $(a/R)^2$, which is approximately several hundredths at ordinary impurity concentrations.

Conclusions

1. The interaction of impurity luminescence centers with the crystal lattice leads a) to a spreading of the absorption and luminescence lines into bands whose width increases with increasing temperature, and b) to a shift of the luminescence bands toward longer wavelengths relative to the absorption band (the so-called Stokes shift).

2. The Stokes shift and the width of the optical impurity band are due to the same physical cause: a change in the forces acting between the impurity atom and its environment. If electrons of the inner shells make the transition, then it is accompanied by practically no change in these forces; this leads simultaneously to the absence of Stokes shift and a small width of the optical band corresponding to the given electronic transition.

3. Not only optical but also nonradiative transitions are possible between the impurity electronic terms. Nonradiative transitions between the levels of the excited and ground states cause a reduction of the luminescence yield, while such transitions between different levels of the excited states lead to a situation in which the absorption of light in different bands can be accompanied by emission in the same luminescence band.

4. The probability of nonradiative transitions increases exponentially with increasing temperature, while at a given temperature it increases exponentially with decreasing distance between the levels. The probability of a nonradiative transition between two close levels usually exceeds the probability of an optical transition.

5. Nonradiative transfer of excitation energy between impurity atoms is possible even for a low impurity concentration.

Chapter II

THE THRESHOLD AND OUTPUT POWER OF A LASER

In the present chapter the threshold pump power and the output power of a laser will be considered from the point of view of the energy relationships without taking into account the spatial nonuniformity of the active medium, which arises from the discrete nature of the spectrum of the laser. This nonuniformity does not lead to any considerable change in the output power (Chaps. IV and V). Since the nonuniformity of the population inversion tends to vanish as the pump power is reduced to the threshold value, it does not effect the threshold pump power.

The treatment is applicable regardless of the mode of operation of the laser as long as the pump duration exceeds the spontaneous emission time of the active centers.

§ 4. The Effective Absorption Coefficient

Let us consider the generation of a light wave in an optically resonant cavity (i.e., the space between two mirrors) which is filled with an active medium. Initially, we shall neglect diffraction phenomena at the boundary of the reflectors and the resulting leakage of light energy from the cavity (diffraction radiation losses will be investigated in subsequent chapters). The light that propagates parallel to the axis of the resonator is amplified by the active medium. The medium has an impurity absorption coefficient (1.12) and it is necessary to take into account the absorption coefficient $\varkappa_0$ of the host material as well. Moreover, at each mirror the light loses a fraction of its energy equal to $1 - r$ (r is the reflectivity of the end mirrors). If these losses are small, then it is possible hypothetically to distribute them over the entire space between mirrors and to consider them by adding a component

$(1 - r)/l$ to the absorption coefficient.† Thus, the interaction of the radiation with the active medium in the cavity is described by the effective absorption coefficient‡ [59, 124]

$$\varkappa(\omega) = \varkappa_1 + \frac{4\pi^2 d^2 \omega}{3\hbar c n(\omega)} [n_1 P_1(\omega) - n_2 P_2(\omega)], \tag{4.1}$$

where

$$\varkappa_1 = \varkappa_0 + \frac{(1 - r)}{l}. \tag{4.2}$$

To simplify the notation, we assume that the matrix element of the dipole moment $\vec{d}_{12}$ is real and drop the subscripts 1, 2; d^2 is defined as a quantity which is averaged over a volume that is assumed to be macroscopically uniform.

Equation (4.1) was written for an active medium in which all the directions of the dipole moment $\vec{d}$ are equiprobable; however, if, for example, all the dipole moments are oriented in the same direction, perpendicular to the optic axis of the laser, then it is necessary to multiply d^2 by 3 in this equation.

We shall use the subscripts 2 and 1 to designate impurity electronic levels between which a working transition exists which is accompanied by generation. Usually there are still other impurity levels which interact with the working levels via nonradiative transitions and are used to reach the predominant population of level 2 (compared with that of level 1), which is required for generation. The form of the effective absorption coefficient depends on the level diagram. Let us examine the following diagrams.

1. The Four-Level Diagram (for example, the rare-earth elements). The 0 level (Fig. 7a) corresponds to the ground state; the pumping is achieved by illuminating the sample in the absorption band corresponding to the 0 → 3 transition. When the pump consists of a blackbody, this band must be fairly broad for the process to be efficient. The 3 → 2 transition occurs nonradia-

† In the case of mirrors having different reflection coefficients, r should be understood to mean their arithmetic mean.

‡ In a laser with external reflectors the distance l between mirrors exceeds the length l_0 of the active sample. In this case the conventional absorption coefficients of the material which appear in Eq. (4.1) (namely, the second term and $\varkappa_0$) must be multiplied by l_0/l.

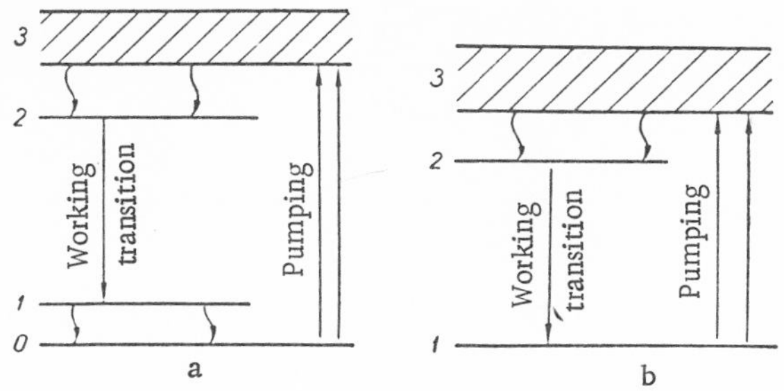

Fig. 7. Level diagram for the operation of a laser with four (a) and three (b) levels. The straight arrows indicate the optical transitions, while the wavy arrows indicate the nonradiative transitions.

tively; the probability of this transition is considerable, since the distance between the 3 and 2 levels is small (see § 2). The 2 → 1 transition, as we have already stated, is a working transition. The transition between the close 1 and 0 levels is achieved nonradiatively with a very high probability. The electrons of the inner shells participate in the working transition, so that the luminescence band 2 → 1 is sufficiently narrow (see § 1), unlike the absorption band 0 → 3, which serves for pumping. These two bands do not intersect, since absorption in the 0 → 3 band does not affect the effective absorption coefficient; this coefficient interests us only within the limits of the luminescence band 2 → 1. Due to the nonradiative transition 1 → 0, the level 1 is practically unpopulated; therefore, the absorption connected with the 1 → 2 transition is similarly absent. Thus, for the four-level diagram we have

$$\varkappa(\omega) = \varkappa_1 - \frac{4\pi^2 d^2 \omega}{3\hbar c n(\omega)} nP(\omega). \tag{4.3}$$

In order to simplify the notation, we drop the subscript 2 which applies to the upper working level.

2. The Three-Level Diagram (one example is ruby). In this case level 1 corresponds to the ground state; the pumping is achieved in a broad absorption band 1 → 3 and is accompanied by nonradiative transitions between the closely spaced levels 3 and 2 (Fig. 7b). The working transition 2 → 1 corresponds to a narrow luminescence band which coincides in position and shape with the absorption band 1 → 2. In other words, the Stokes shift is practi-

cally absent.† As was shown in § 1, the small half-width of the optical band $1 \rightarrow 2$ and the absence of Stokes shift (which is related to the half-width by Eq. (1.9)) are due to the same physical cause: the fact that internal electrons, which do not affect the relationship between the impurity atom and its nearest neighbors, participate in the working transition.

Due to the high probability of the nonradiative $3 \rightarrow 2$ transition, it is possible to place $n_1 + n_2 = n_0$, where n_0 is the total number of impurity luminescence centers per unit volume. Considering this and using the substitutions $P_1(\omega) = P_2(\omega) = P(\omega)$, $n_2 = n$, we have the following result for the three-level diagram:

$$\varkappa(\omega) = \varkappa_1 + \frac{4\pi^2 d^2 \omega}{3\hbar c n(\omega)} (n_0 - 2n) P(\omega). \tag{4.4}$$

3. The Two-Level Diagram in the Presence of Stokes Shift. Assume that the optical bands $1 \rightarrow 2$ and $2 \rightarrow 1$ have a considerable Stokes shift (and therefore a large half-width). Neglecting absorption in the vicinity of the luminescence maximum, we obtain the same equation (4.3) as we obtained in the case of the four-level diagram for the effective absorption coefficient.‡

§5. The Threshold Pump Power

Let us now examine an active medium placed between plane-parallel mirrors and excited by optical pumping (the reasoning remains valid for concave reflectors also). For a low pump power N the effective absorption coefficient is positive, and the radiation in the resonator is damped; therefore, there is no generation. If the pump power is increased continuously, then for a certain value $N = N^*$ of it the effective absorption coefficient¶ $\varkappa(\omega_0)$ vanishes, and for a further negligibly small increase in pumping it acquires

† We should not confuse Stokes shift, which is characterized by the relative positions of the optical bands that are connected with the same electronic transition, with the shift of the pumping band $1 \rightarrow 3$ relative to the luminescence band $2 \rightarrow 1$.

‡ Hereafter, in speaking of the four-level diagram, we shall also have in mind the two-level diagram.

¶ It is easy to see that for all of the level diagrams examined above, the effective absorption coefficient $\varkappa(\omega)$ reaches a minimum at the point $\omega = \omega_0$, where ω_0 is the frequency corresponding to the maximum of the luminescence band (more precisely, the maximum of the function $P(\omega)$ which stipulates the shape of this band).

a negative value having a small absolute magnitude (ω_0 designates the frequency corresponding to the maximum of the luminescence band 2 → 1). Then radiation at the frequency ω_0, which propagates parallel to the optic axis, starts to increase exponentially with time, and therefore it will predominate sharply over radiation having a different frequency and a different direction of propagation. This is a characteristic attribute of the stimulated emission.

Thus, the generation condition has the form

$$\varkappa(\omega_0) = 0. \tag{5.1}$$

The quantity n = n*, which satisfies Eq. (5.1), is the thresold value of the volume concentration of the excited luminescence centers; the pump power required to maintain this value of n is the threshold power. Obviously, the threshold pump power is equal to the spontaneous-emission power of an active medium having the threshold value n (in order to avoid encumbering the formulas, we shall assume that the luminescence energy yield is equal to unity). Taking account of Eq. (1.3), we find the threshold pump power absorbed per unit volume of the active medium [59]:

$$N^* = \frac{4}{3} \cdot \frac{\omega_0^4 n(\omega_0) d^2}{c^3} n^*. \tag{5.2}$$

Setting Eq. (4.3) for $\varkappa(\omega)$ equal to zero, we calculate n* for the four-level diagram. Substituting it into Eq. (5.2), we find the threshold power absorbed per unit volume for a four-level laser:

$$N^* = \frac{\hbar\omega_0^3}{\pi^2 v^2 P(\omega_0)} \left(\varkappa_0 + \frac{1-r}{l_0}\right) \left(v = \frac{c}{n(\omega_0)}\right). \tag{5.3}$$

Here l_0 is the length of the active sample, which can be shorter than the resonator length l. Analogously, making use of Eq. (4.4), we find the threshold pump power for a three-level laser:

$$N^* = \frac{\hbar\omega_0^3}{2\pi^2 v^2 P(\omega_0)} \left[\mu(\omega_0) + \varkappa_0 + \frac{1-r}{l_0}\right], \tag{5.4}$$

where μ is the impurity light absorption coefficient of the unexcited active medium in the 1 → 2 band. Note that for a Lorentz luminescence band this equation coincides with the well-known equation derived by Schawlow and Townes [101], who used different terminology and assumed that all of the optical bands were Lorentz bands.

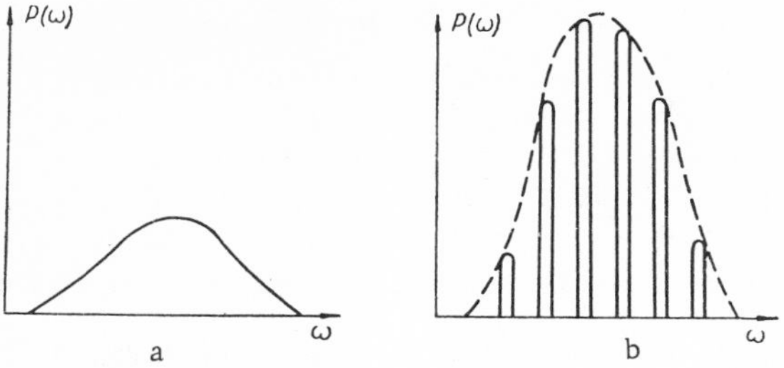

Fig. 8. Two luminescence bands (a and b) having an identical shape but differing values of the maximum ordinate of the curve $P(\omega_0)$.

If the shape of the luminescence band is represented by a smooth bell-shaped curve, then $P(\omega_0) \sim 1/\Delta\omega$; therefore, as the luminescence band becomes narrower, the threshold power decreases. If we assume that the function $P(\omega)$ is Gaussian, then we have the following result for a four-level diagram:

$$N^* = 0.108 \frac{\hbar\omega_0^3 \Delta\omega}{v^2} \left(\varkappa_0 + \frac{1-r}{l_0}\right) \left(v = \frac{c}{n(\omega_0)}\right), \qquad (5.5)$$

while for a three-level diagram we have

$$N^* = 0.054 \frac{\hbar\omega_0^3 \Delta\omega}{v^2} \left[\mu(\omega_0) + \varkappa_0 + \frac{1-r}{l_0}\right]. \qquad (5.6)$$

However, if the luminescence band has a vibrational structure† (Fig. 8b), then $P(\omega_0)$ can exceed the reciprocal of the magnitude of the luminescence-band half-width considerably; correspondingly, the threshold pump power will be smaller than in the case of a smooth $P(\omega)$ curve having the same half-width. Thus, the curve shown in Fig. 8b has a much greater maximum ordinate $P(\omega_0)$ than the curve shown in Fig. 8a because of the vibrational structure.

So far we have assumed that the reflectivity r is independent of frequency. Let us now examine the case in which the end mirrors have selective reflection in a narrow frequency band whose width is smaller than the width of the luminescence band. Then the working frequency is the frequency ω_r that corresponds to the reflection maximum, and in Eqs. (5.3) and (5.4) the function $P(\omega)$ must be

† The vibrational structure of optical impurity bands can be resolved in those cases when their shape is determined predominantly by local vibrations of the impurity molecule rather than by the interaction of this molecule with lattice vibrations.

taken at the point $\omega = \omega_r$ (Eqs. (5.5) and (5.6) are inapplicable in this case). Thus, the working frequency can be stipulated using selective reflectors; here we find that $P(\omega_r) \sim P(\omega_0)$ if this frequency ω_r does not depart beyond the limits of the luminescence band, and the threshold power does not increase very strongly.

Hereafter we shall assume that r does not depend on frequency.

From Eq. (5.4) it is evident that for the three-level diagram the threshold power decreases in proportion to the total concentration n_0 of luminescence centers as the absorption coefficient $\mu(\omega_0)$ decreases. However, this concentration cannot be made too small. In fact, even if all luminescence centers are transferred to an excited state (i.e., $n = n_0$), it follows that the effective absorption coefficient (4.3) or (4.4) can vanish only for the condition

$$n_0 > n_{0\min} \equiv \frac{0.108 T \omega_0^2 \Delta\omega}{v^2} \left[\varkappa_0 + \frac{1-r}{l_0} \right] \tag{5.7}$$

(T is the time required for the spontaneous optical transition 2 → 1). If n_0 exceeds $n_{0\min}$ only slightly, it follows that the equation for the threshold power of a three-level laser does not differ from the corresponding equation for the four-level diagram.

In considering the problem of optimal concentration of a luminescent impurity, we recall that N does not represent the total luminosity of the lamps used for pumping but merely that portion of the luminosity which is absorbed in the pump band of the active medium. Therefore, the threshold value of the luminosity of the lamps (referred to a unit volume of the active medium) is equal to N^*/η, where η is the light-energy utilization coefficient (usually, η is several percent). In the case of the four-level diagram the concentration of the luminescent impurity atoms does not affect N^*, and therefore it must be chosen exclusively on the basis of consideration of optimal pump conditions, which provide for the maximum η. In the case of the three-level diagram, however, both N^* and η (for $n_0 = n_{0\min}$ we have $\eta = 0$, since all of the impurity centers are transferred to an excited state and 1 → 3 absorption is absent) increase with increasing impurity concentration n_0; therefore, the impurity concentration must be chosen with allowance for the two competing factors indicated.

All of the equations given above for the threshold power were written for an optically isotropic active medium in which all orientations of the dipole moment are equally probable (in particular, for a crystal having cubic symmetry). However, if the dipole moments of all the impurity centers are directed in one direction at the angle φ relative to the optic axis of the laser, then the threshold power N^* is equal to the value of N^* for the isotropic case, multiplied by the quantity $1/(3 \sin^2 \varphi)$.

It is common practice to assume that generation is possible only when the population of the levels is inverted (i.e., for the conditions $n \equiv n_2 > n_1$). From the above it is clear that this is valid for three- and four-level diagrams but not for two-level diagrams with Stokes shift [18]. In fact, for sufficiently small $\varkappa_1$ the effective absorption coefficient for the two-level diagram† (4.3) can vanish even for $n_2 \equiv n < n_0/2$ (i.e., for $n_2 < n_1$). Physically, this is connected with the fact that because of Stokes shift the luminescence centers which are in the ground state do not absorb the generated light.

The apparent paradox in the fact that generation is also possible without a population inversion of the electronic working terms can easily be explained if we remember that each of these terms corresponds to a system of vibrational levels. In order to obtain generation it is sufficient to have a population inversion of the lowest vibrational levels of the upper electronic term relative to certain vibrational levels of the lower terms; this population can easily be made compatible with the inequality $n_2 < n_1$, where n_2 and n_1 are the resultant occupancies of the system of vibrational levels corresponding to the upper or lower term.

However, the large spectral luminescence width which unavoidably accompanies a considerable Stokes shift (see § 1) and which leads to an excessively high generation threshold creates serious difficulties for the experimental implementation of a two-level laser with Stokes shift. Therefore, in practice the condition of inverse population of the working electronic terms remains in force.

§ 6. The Output Radiation Power of a Laser

Assume that the pump power increases continuously, beginning at zero. As we have seen, for $N < N^*$, where N^* is the thresh-

† We recall that in the case of a two-level diagram with Stokes shift the equations for the effective absorption coefficient and the threshold power have the same form as they do for a four-level diagram.

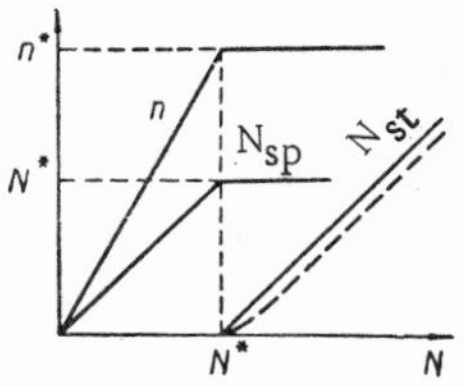

Fig. 9. Graph of the dependence of the number of excited atoms n and the amounts of power corresponding to spontaneous (N_{sp}) and stimulated (N_{st}) emission on the pump power N.

old power, the effective absorption coefficient is positive, and there is no generation. When the pump power N reaches the value N^* and the number of excited atoms reaches the threshold value n^*, the effective absorption coefficient vanishes, and generation begins. For a further increase in the pumping, n remains practically equal to the threshold value.† In fact, if n were to exceed n^* slightly, then the effective absorption coefficient would be negative, and the light energy in the cavity would begin to increase exponentially with time. This would lead to intense stimulated emission from excited atoms, which would continue until n decreased to the threshold value. Therefore, the dependence of n on the pumping has a break at the point N^*, as shown in Fig. 9.

The pump power absorbed by the luminescence centers is converted into the power of spontaneous emission N_{sp} and the power of stimulated emission N_{st}, i.e.,

$$N = N_{sp} + N_{st} \tag{6.1}$$

(for simplicity we assume that the luminescence energy yield is equal to unity). Until the generation threshold is reached, the second term in (6.1) is equal to zero, and the entire pump power is converted into spontaneous emission. After the threshold n has been reached, n remains equal to n^*, and the power of spontaneous emission remains equal to the threshold power N^*; the power of directional stimulated emission is equal to the difference $N - N^*$ (Fig. 9).

On the experimental curves the break in the dependence of N_{st} on N is smoothed somewhat (as shown in Fig. 9 by the dashed line). This is connected with the nonuniform distribution of the pump intensity over the cross section of the active sample. A cylindrical sample serves as a lens which focuses the pumping

† In the relaxation mode (Chap. VII) insignificant oscillations take place near the threshold value also.

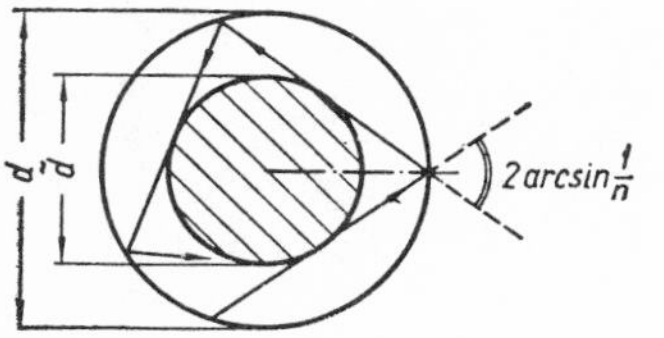

Fig. 10. Generation of the ring rays.

light close to the axis of the cylinder [95]; for a slight excess above the threshold only a small central portion of the sample has negative absorption. With a further increase in the pumping the value of N_{st} increases more slowly than it does for a large excess above the threshold.

Moreover, the entire cross section of the sample usually does not generate radiation even for a pump power considerably above the threshold. As a rule, laser rods have a cylindrical shape and a smooth side; because of the total internal reflection from the side, the rays can move in the plane of the cross section without emerging from the sample (Fig. 10) and without undergoing losses during reflection from the end mirrors. Such ring rays (so-called "whispering modes") are subject to more favorable conditions than the generated radiation, and the latter cannot compete with it. Therefore, the directed radiation is generated only in that region of the cross section which is inaccessible to the ring rays. From Fig. 10 it is obvious that the diameter of the working portion of the cross section has the form

$$\tilde{d} = \frac{d}{n(\omega_0)}, \tag{6.2}$$

where d is the rod diameter; $n(\omega_0)$ is the refractive index. Therefore, due to the ring rays the output power of the laser decreases approximately by a factor of $[n(\omega_0)]^2$, and the working portion of the cross section (shaded in the figure) decreases by a factor of $n(\omega_0)$. Equations (6.2) have been substantiated experimentally [95, 96].

In order to eliminate the ring rays the lateral surface of the sample is given a matte surface, and grooves are cut in it. A more detailed description of off-axial rays and methods of eliminating them is contained in [26, 48].

Let us calculate the output power of a laser while taking into account actual experimental conditions [32]. The radiation is extracted from the cavity only through one mirror, which we shall designate by the subscript 1; this mirror is made more transparent than the mirror 2. Let us consider all possible losses of light energy in the resonator. The mirrors have reflectivities r_1 and r_2, which differ from unity; here $1 - r_1 = A_1 + \Pi_1$, $1 - r_2 = A_2 + \Pi_2$, where A_1 and A_2 are the absorption coefficients of mirrors 1 and 2, while Π_1 and Π_2 are their transmissivities. Further, a single pass of the light through the active rod is accompanied by a loss of a fraction of its intensity equal to $\varkappa_0 l_0$, where $\varkappa_0$ is the absorption coefficient of the host substance. Finally, it is necessary to consider the losses connected with the deflection of the direction of light propagation from the optic axis during the passage of light through an optically imperfect active rod and during reflection from reflectors which have not been fabricated accurately enough. We use $\tilde{A}$ to designate the relative magnitude of these losses for a single pass of the light through the resonator.

During a round trip through the resonator, the light passes through the active medium twice and is reflected once from each mirror. Thus, the resultant magnitude of the relative light losses for a single pass of the light through the resonator is equal to

$$A + \frac{1}{2}\Pi_1, \tag{6.3}$$

where

$$A = \frac{1}{2}(A_1 + A_2 + \Pi_2) + \tilde{A} + \varkappa_0 l_0 \tag{6.4}$$

is the sum of all the losses with the exception of the term $\Pi_1/2$, which is connected with the transmissivity of the mirror 1. Below we shall assume that A is a constant and Π_1 is varied.

Using the notation introduced above, it is not difficult to write the output power of the laser (i.e., the luminosity extracted via the mirror 1):

$$W = (N - N^*)\frac{\Pi}{2A + \Pi_1} \tag{6.5}$$

(the output power, just as the pump power, is referred to a unit volume of the active medium). The first factor in this equation represents the stimulated-emission power in the resonator, while the second represents the ratio between the intensity losses connected

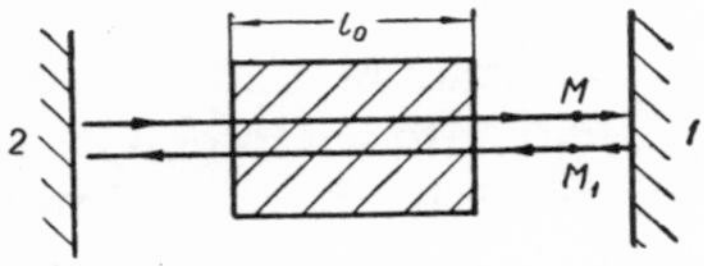

Fig. 11. Diagram for derivation of the equation for the threshold power in the presence of large light-energy losses.

with the exit of the radiation and the resultant magnitude of the stimulated-emission losses in the resonator.†

It is necessary to substitute the expression for the threshold power N* into Eq. (6.5). However, the equations for the threshold power obtained in the previous section become insufficiently accurate if the mirror 1 used to extract the radiation from the resonator has a transmissivity comparable with unity. Rigorously speaking, the light losses which accompany the reflection from such a mirror cannot be considered by adding the term $(1 - r)/l$ to the absorption coefficient.

We shall calculate the threshold pump power anew without using the effective absorption coefficient; for the sake of brevity we shall restrict ourselves to the case of a four- or two-level laser (the case of a three-level diagram can be examined in a completely analogous manner).

The threshold power is determined from the condition requiring that a ray emanating from a certain point M (Fig. 11) returns to this point after a round trip through the resonator with its previous intensity. If the initial intensity is taken to be unity, then the intensity is equal to $r_1 = 1 - A_1 - \Pi_1$ at the point M_1 after reflection from the mirror 1; after passage through the sample, the intensity increases in the ratio $\exp[-(\tilde{\mu}(w_0) + \varkappa_0)l_0]$, where $\tilde{\mu}$ is the coefficient (negative) of impurity absorption. Reasoning in a similar manner we find the intensity of the ray which has returned to the point M after a round trip through the resonator:

$$(1 - A_1 - \Pi_1)(1 - A_2 - \Pi_2)(1 - \tilde{A})^2 \exp\{-2l_0[\tilde{\mu}(\omega_0) + \varkappa_0]\}.$$

Setting this expression equal to unity, we find the quantity $\tilde{\mu}(\omega_0)$,

† For simplicity, we have examined the case of unpolarized or completely polarized radiation here. However, in the general case the interaction of electromagnetic fields having different polarizations can lead to a certain violation of the linear dependence of W on N (see Chapter IX for greater detail).

which is proportional to the number of excited atoms (and therefore to the threshold pump power) for a four-level diagram. Thus,

$$N^* = C[-\ln(1-A_2-\Pi_2) - \ln(1-A_1-\Pi_1) - 2\ln(1-\widetilde{A}) + 2\varkappa_0 l_0], \quad (6.6)$$

where C is a constant which shall be determined below.

Usually, all of the components of the light losses, with the exception of the transmissivity of the mirror 1 through which the light is extracted from the resonator, are small compared to unity. Expanding Eq. (6.6) into a series in the small losses, we find the following results with an accuracy of up to linear terms:

$$N^* = C[2A + \Pi_1 + \varphi(A_1 + \Pi_1)], \quad (6.7)$$

where

$$\varphi(x) = -\ln(1-x) - x = \frac{x^2}{2} + \ldots \quad (6.8)$$

is a function whose graph is shown in Fig. 12. If $A_1 + \Pi_1 \ll 1$ (i.e., the losses on mirror 1 are small), then the latter term in Eq. (6.7), which is quadratic in the losses, can be dropped, and this equation coincides with Eq. (5.3), which was obtained using the effective absorption coefficient (it is possible to determine the constant C from a comparison of these two equations).

Thus, if each component of the light losses (but not necessarily their sum!) is small compared with unity, then the dependence of the threshold power on the resultant losses is linear. However, if the losses on one of the mirrors are comparable with unity, then the term φ in Eq. (6.7) represents a correction which is usually

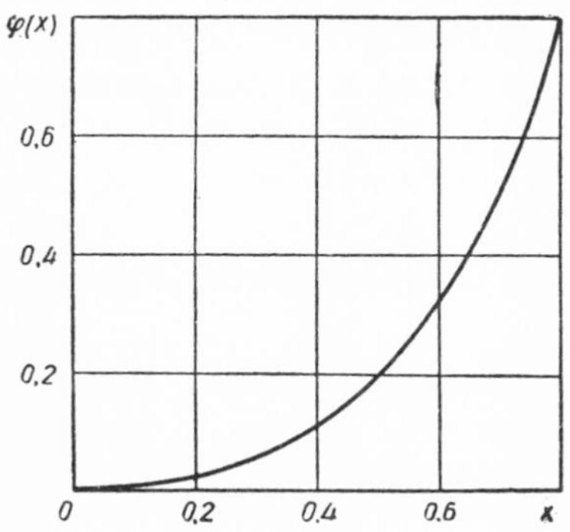

Fig. 12. Graph of the function $\varphi(x)$.

small and which describes the deviation of this dependence from a linear relationship.

Substituting (6.7) into Eq. (6.5), we find the output power

$$W = \{N - C[2A + \Pi_1 + \varphi(A_1 + \Pi_1)]\} \frac{\Pi_1}{2A + \Pi_1} . \qquad (6.9)$$

Let us examine the dependence of the output power on the transmissivity of the mirror 1 through which the radiation is emitted. First we express the constant C in terms of the parameters which can be determined directly from experiments. Assuming in Eq. (6.7) that $\Pi_1 = 0$, $\varphi = A_1^2/2 \cong 0$, we find $C = N_0^*/2A$, where N_0^* is the threshold pump power for $\Pi_1 = 0$. Substituting into Eq. (6.9), we reduce the expression for the output power to its final form:

$$W = N_0^* \left[\xi \frac{\Pi_1}{2A + \Pi_1} - \frac{\Pi_1}{2A} - \frac{\Pi_1 \varphi (A_1 + \Pi_1)}{2A(2A + \Pi_1)} \right], \qquad (6.10)$$

where $\xi = N/N_0^*$ is the ratio between the pump power and its threshold value for $\Pi_1 = 0$.

Equation (6.10) has a maximum with respect to Π_1, since for $\Pi_1 = 0$ it vanishes, while for $\Pi_1 \to \infty$ it decreases with increasing Π_1. Setting the derivative of W with respect to Π_1 equal to zero, we find the optimum transmissivity which corresponds to the maximum output power for a given pump power:

$$\Pi_{max} = 2A(\sqrt{\xi} - 1) - \delta\Pi_1, \qquad (6.11)$$

where $\delta\Pi_1$ designates the small correction

$$\delta\Pi_1 = 4A\sqrt{\xi} - \sqrt{4\xi A^2 - 2A\varphi(A_1 + \Pi_1) - \frac{\Pi_1(\Pi_1 + A_1)(\Pi_1 + 2A)}{1 - A_1 - \Pi_1}}$$

This correction can be calculated for $\Pi_1 = 2A(\sqrt{\xi} - 1)$. If $2A(\sqrt{\xi} - 1) \ll 1$, then this correction can be neglected, and then Eq. (6.11) coincides with the well-known formula for the optimum transmissivity [90].

Equation (6.11) has a clear meaning. The optimal value of the transmissivity Π_1 is determined by the competition between two relationships: the fraction of light energy emitted from the resonator increases with increasing Π_1, but at the same time the threshold pump power increases, thus decreasing the stimulated-emission power in the resonator (this power is equal to $N - N^*$). This increase of the threshold plays a role which diminishes as the pumping increases; therefore, $\Pi_{1\,max}$ increases with an increase in pump power.

All of the quantities appearing in Eqs. (6.10) and (6.11) are known parameters of the reflectors, the active medium, and the pumping. The quantity $\tilde{A}$ is an exception (i.e., the magnitude of the relative losses connected with the diffraction of the light by the optical system). As we shall show in Chaper VI, these losses can be reduced by means of a collecting lens placed inside the cavity. This technique can be used for the experimental determination of $\tilde{A}$; in this it is necessary to consider the fact that the lens introduces new relative losses $2r_l$, where r_l is the reflectivity of the glass from which the lens is fabricated. Writing Eq. (6.7) for a resonator with and without a lens and dividing the first equation term-by-term by the second we obtain the equation for $\tilde{A}$, from which we find

$$\tilde{A} = [A_1 + A_2 + \Pi_1 + \Pi_2 + 2\varkappa_0 l_0 + \varphi(A_1 + \Pi_1)]\left(\frac{N^*}{N_l^*} - 1\right) + 2r_l\frac{N^*}{N_l^*} \quad (6.12)$$

where N_l^* is the threshold power for the laser with the lens.

As an example, we shall present the results of an experimental investigation of the dependence of the output power on transmissivity for a four-level laser based on neodymium-glass [32]. The experiments were carried out with a laser having the following loss components: $\tilde{A} = 0.13$, $\varkappa_0 l_0 = 0.10$, $A_1/2 = A_2/2 = 0.02$, $\Pi_2/2 = 0.07$. The curve for the dependence of W on Π_1, plotted from these data in accordance with Eq. (6.10), is shown in Fig. 13; the open circles and crosses show the experimental values of the output power obtained with reflectors of different types.

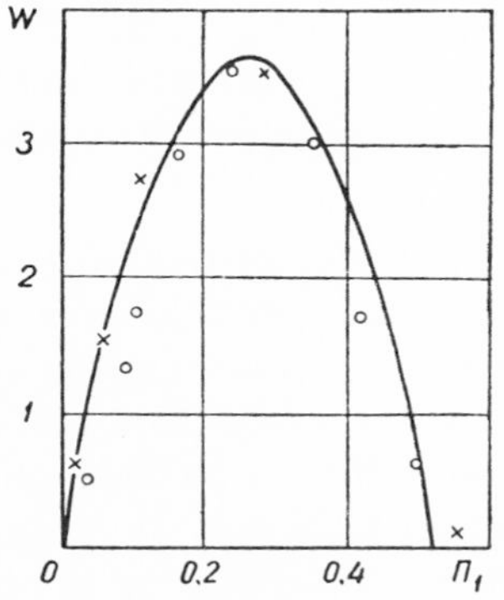

Fig. 13. Dependence of the output energy on the transmissivity of one of the reflectors.

Conclusions

1. As the pump power N increases up to the threshold value N^*, the number of excited atoms n increases with an increase in the pumping; there is no stimulated emission. At $N > N^*$ generation of radiation with intensity proportional to the difference $N - N^*$ takes place, while the number of excited atoms remains approximately equal to its threshold value, which corresponds to the threshold pump power.

2. The threshold power decreases with a decrease of the light-energy losses, the luminescence bandwidth, and the light frequency.

3. For a three-level laser an optimum concentration of impurity luminescence centers exists which corresponds to the minimum threshold pump power.

4. The imperfection of the optical system leads to additional losses of light energy in the cavity, which cannot be neglected in the general case. This loss component can be determined by means of a positive lens which is placed inside the cavity.

5. For a given pump power an optimum mirror transmission exists which corresponds to the maximum laser output power. The optimum transmissivity increases with increasing pumping and light losses in the cavity.

6. In a cylindrical active sample having a polished surface, only the central part of the cross section generates radiation because of the existence of ring rays. Ring rays can be eliminated by the appropriate treatment of the lateral surface.

Chapter III

THE NATURAL OSCILLATIONS OF AN IDEAL RESONATOR (LINEAR THEORY)

For the sake of brevity, we shall define a resonator as an optical cavity in which there are no light-energy losses (with the exception of small diffraction losses) and no pumping; i.e., there is actually no active medium. The electromagnetic field in an ideal resonator is described by a linear wave equation. As we shall show in the subsequent chapters, in the case of a practical resonator the wave equation is nonlinear, and the nonlinearity vanishes as the light losses and excess above the threshold pump power tend to vanish. Thus, an ideal resonator is the limiting case of a practical resonator. The ideal resonator is approximated to varying degrees by the experimental conditions.†

§7. Oscillating Modes of a Closed Resonator

We begin by examining a closed ideal resonator (i.e., a resonator with perfectly reflecting walls).† In such a resonator there are no losses connected with the radiation of light outward (these losses are also called diffraction losses, since they are caused by the diffraction of light by the end mirrors). A closed resonator can be achieved experimentally to a high approximation if the reflectors have a reflectivity close to unity.¶ They can be vacuum-deposited onto the faces of a rod having a smooth surface. The rays which form an angle with the optic axis which is not too large under-

† Gas lasers are usually close to an ideal resonator because of the weak pumping and the absence of losses in the material.

‡ An excellent treatment of the subject of cavity modes can be found in [118].

¶ Under these conditions the shape of the rod cross section must be chosen in such a way that ring rays are eliminated (§6).

go complete internal reflection from the lateral surface, and the electromagnetic field does not exit from the confines of the resonator. A more ingenious experimental version of a closed resonator is an active sample fabricated in the form of a closed torus [99].

For simplicity, let us consider a closed resonator in the shape of a rectangular parallelepiped having a length l and cross-section sides $2a$ and $2b$. Let us place the origin at the center of the end face of the parallelepiped and direct the z axis along the optic axis of the resonator, while the x and y axes are directed parallel to the sides of the cross section. We write the wave equation for a homogeneous dielectric medium:

$$\Delta E + k_0^2 E = 0, \quad k_0^2 = \frac{\omega^2 \varepsilon_0}{c^2} \equiv \frac{\omega^2}{v^2}. \tag{7.1}$$

Here ω is the light frequency; k_0 is its wave vector; ε_0 is the real part of the dielectric constant of the medium; v is the velocity of light in the medium. For simplicity, let us examine the case of plane-polarized light and let us use E to designate the modulus of the electric-field vector. Equation (7.1) should be solved for the boundary conditions

$$E\,|_{z=0} = E\,|_{z=l} = E\,|_{x=\pm a} = E\,|_{y=\pm b} = 0. \tag{7.2}$$

It is easy to verify the fact that the solution has the form†

$$E = E_0 \sin \frac{\pi m_3 z}{l} \cos \frac{\pi y}{b}\left(m_2 + \frac{1}{2}\right) \cos \frac{\pi x}{a}\left(m_1 + \frac{1}{2}\right), \tag{7.3}$$

where the integers m_1, m_2, m_3 are interrelated by the equation

$$\frac{m_1^2}{a^2} + \frac{m_2^2}{b^2} + \frac{m_3^2}{l^2} = \frac{k_0^2}{\pi^2} \equiv \frac{\omega^2 \varepsilon_0}{\pi^2 c^2} \equiv \frac{\omega^2}{\pi^2 v^2}. \tag{7.4}$$

This equation stipulates the spectrum of the natural frequencies of the closed resonator; all of the frequencies are real, which corresponds to the absence of diffraction losses.

The solution (7.3) describes three systems of standing waves in the direction of the x, y, z axes. The index m_3, which is equal to

† In order to simplify the notation we shall consider transverse modes having only even numbers (counting the fundamental mode as the zero mode). Odd modes can be treated in a completely analogous manner, but their consideration will not yield anything new.

the number of half-waves which will fit within the gap between mirrors, is called the longitudinal mode index, whereas the indices m_1 and m_2 are called the transverse modes indices†

If we restrict our analysis to transverse modes having small indices, then the dependence of the frequency on the indices of the modes takes the form

$$\omega(m_1, m_2, m_3) = \pi v \left(\frac{m_3}{l} + \frac{l}{2a^2} \cdot \frac{m_1^2}{m_3} + \frac{l}{2b^2} \cdot \frac{m_2^2}{m_3} \right). \quad (7.5)$$

Since usually $m_3 \sim 10^5$, it is evident that the transverse modes having low indices are spaced considerably more closely than the longitudinal modes.

If in the solution (7.3) we represent each sine as the sum of two traveling waves, then it is not difficult to verify the proposition that the angle between the direction of light propagation and the optic axis in the xz plane is equal to the diffraction angles $\lambda / 2a$ multiplied by the transverse-mode index m_1; similarly, the angular divergence of the light in the yz plane is equal to $\lambda m_2/2b$. Since no constraints are imposed on the indices of the modes (with the exception of equations (7.4)), it follows that the angular spread of the radiation in a completely closed resonator is relatively large and can be comparable with unity. Moreover, the large angular spread of the radiation is also obvious without recourse to an examination of the modes. In fact, axial rays do not offer any advantages over oblique rays, since the latter do not undergo any losses during reflection from the resonator walls.

Below it shall be shown that in the case of an open resonator the diffraction losses are considerably greater for oblique rays than they are for axial waves, so that the latter sharply predominate over the former.

§8. The Modes of an Open Resonator

It is common practice to use resonators which are open in varying degree. Different versions of such resonators are shown in Fig. 14.

† The natural oscillations of the electromagnetic field in the resonator are designated by the symbol $TEM_{m_1 m_2 m_3}$.

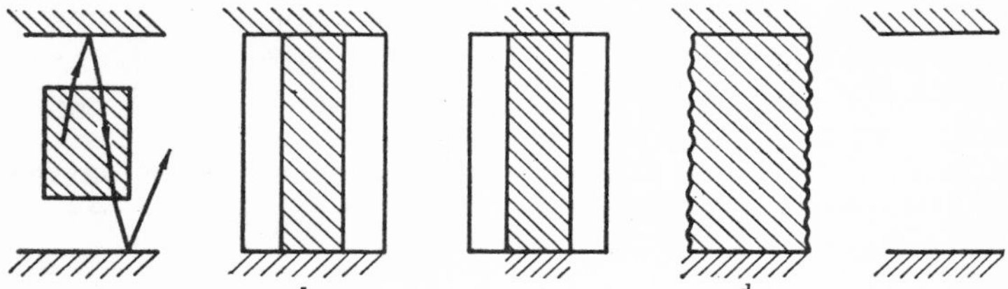

Fig. 14. Open resonators of different types: a) resonator with external reflectors; b) resonator with a polished cylindrical active rod; c) active rod with a nonactive coaxial cladding; d) active rod with a matte lateral surface; e) empty open resonator.

In a resonator with external reflectors (Fig. 14a) a portion of the space remains open; this leads to the sharp predominance of axial rays, since oblique rays exit from the resonator after several passes and are lost (this is shown in the figure).

As has already been noted in § 6, in a circular rod having a smooth lateral surface the diameter of the active region is smaller than the rod diameter. The working part of the volume (shaded in Fig. 14b) can be treated as an open resonator. The situation is similar in the case of an active rod cladded with an inactive transparent material having the same refractive index (Fig. 14c). Such cladding is used to eliminate ring rays and to focus the pumping light in the active rod.† In the case shown in Fig. 14 the axial rays predominate substantially over oblique rays, which depart rapidly from the active region. This also derives from consideration of the wave equation (see below).

In order to eliminate ring rays the lateral surface of the rod is sometimes made into a matte surface (Fig. 14d). On collision with the rough surface the rays are scattered and partially exit to the outside. Therefore the rod can be treated as a partially open resonator.

Resonators of the type shown in Figs. 14b through 14d can have external reflectors, as shown in Fig. 14a. Because of this they appear even more open.

---

† Cladding also absorbs ultraviolet light from the pump. This prevents "solarization" of the rod and increases its useful life.

Finally, we can conceive of an open resonator which does not contain any material (Fig. 14e); such a resonator is a close approximation to certain gas lasers.

Open resonators have been investigated in detail in a number of papers [22, 23, 79, 85, 86]; the most modern theory of open resonators is attributable to Vainshtein [22, 23], who examined a resonator of the type "e" in Fig. 14. We shall present a simpler (although somewhat less rigorous) discussion of a resonator of the type shown in Fig. 14b or 14c.

Let us draw the coordinate axes as was done in the previous section. In the active region I (Fig. 15) there are no light losses (with the exception of diffraction losses, which shall be considered below) in accordance with the definition of an ideal resonator; therefore, the dielectric constant is real, and the wave equation retains the form (7.1).

However, in region II, which belongs to the inactive cladding, the light undergoes relative losses β during each pass through the resonator, where β is a certain number of the order of unity.

In order to simplify the treatment we shall assume the reflection from the faces to be complete, while the light losses shall be distributed uniformly over the material and shall be described by the absorption coefficient β / l, $\beta \sim 1$. This coefficient corresponds to the complex dielectric constant

$$\varepsilon = \varepsilon_0 \left(1 + \frac{i\beta}{2k_0 l}\right)^2 = \varepsilon_0 \left(1 + \frac{i\beta}{k_0 l}\right) \qquad (\beta \sim 1) \qquad (8.1)$$

(we neglect the square of the small quantity $1/k_0 l$). The wave equation in the region II differs from Eq. (7.1) solely in the fact that ε_0 must be replaced by the complex dielectric constant (8.1). The solutions of these two equations must make the transition in-

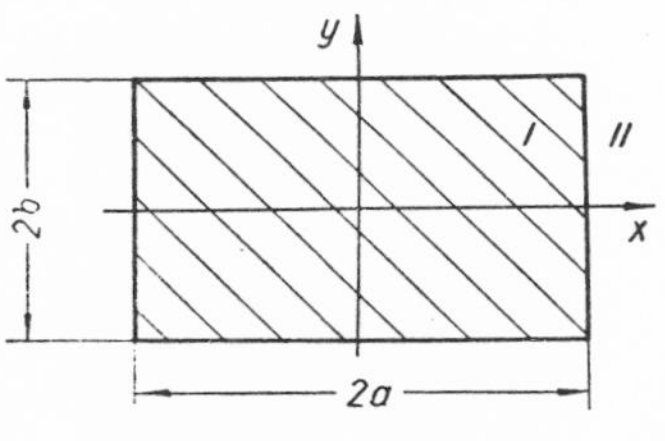

Fig. 15. Cross section of the resonator.

to one another on the boundaries between the regions I and II; the condition for matching these solutions replaces the condition for the vanishing of the field on the lateral surface of the rod, which was used in §7. However, the condition for the vanishing of the field on the surface of the end reflectors remains valid in both regions I and II.

The other difference from the case of a closed resonator resides in the fact that the light energy undergoes diffraction losses, so that the frequency ω is complex.

In the region I the solution of the wave equation has the form

$$E = E_0 \sin k_3 z \cos k_2 y \cos k_1 x, \tag{8.2}$$

where

$$k_3 = \frac{\pi m_3}{l}, \quad k_1^2 + k_2^2 + k_3^2 = k^2 = \frac{\omega^2 \varepsilon_0}{c^2} = \frac{\omega^2}{v^2}. \tag{8.3}$$

In the region II the solution can be written as follows:

$$E = E_1 \sin k_3 z \cos k_2 y e^{-p(x-a)}, \tag{8.4}$$

where

$$k_3^2 + k_2^2 - p^2 = k^2\left(1 + \frac{i\beta}{k_0 l}\right) = \frac{\omega^2}{v^2}\left(1 + \frac{i\beta}{k_0 l}\right), \quad \beta \sim 1, \tag{8.5}$$

here k_2 and k_3 are the same as in the solutions (8.2). We chose the solution in a form which provides for the possibility of matching with the solution (8.2) in the plane† $x = a$. As in § 7, we shall examine only the even transverse modes in order to simplify the notation.

Assuming that the diffraction losses are sufficiently small, we shall seek the wave vector $\vec{k}$ in a form which is close to that used in the case of a closed resonator:

$$k_1 = \frac{\pi}{a}\left(m_1 + \frac{1}{2}\right) + \frac{\xi_1}{a}, \quad k_2 = \frac{\pi}{b}\left(m_2 + \frac{1}{2}\right) + \frac{\xi_2}{b}; \; k_3 = \frac{\pi m_3}{l}, \tag{8.6}$$

where the increments ξ_1 and ξ_2 are small compared to unity.

† The solution (8.4) is applicable near the plane $x = a$; near the other boundaries of the active region of the cross section ($x = -a$, $y = \pm b$) the solution should be sought in a different form. However, in view of the small penetration depth of the field beyond the limits of the active region of the cross section, the solution should be treated only in a very narrow layer (having a thickness of the order of $(l/k_0)^{1/2}$) which is adjacent to the planes $x = \pm a$, $y = \pm b$; this allows the solution to be sought independently near each of these planes.

Comparing (8.3) with (8.5), we find

$$p^2 = k_1^2 - \frac{i\beta k_0}{l}. \tag{8.7}$$

The ratio between the first term in the right side of this equation and the second term is a quantity of the order of $m_1^2 \lambda l / a^2$.

We introduce the small parameter of the theory:

$$D = \frac{\lambda l}{a^2} \ll 1. \tag{8.8}$$

Below we shall see that the smallness of this parameter provides for the smallness of the diffraction losses of the generated radiation. It shall also be shown that transverse modes exist only for the condition

$$m_\perp \ll D^{-\frac{1}{2}}, \tag{8.9}$$

where m_1 is the index of the transverse modes (m_1 or m_2).

Making use of the inequality (8.9), we neglect the first term in Eq. (8.7). Thus, we find p:

$$p = (1 - i)\sqrt{\frac{\beta k_0}{2l}}. \tag{8.10}$$

Setting the solutions (8.2) and (8.4) and their derivatives with respect to x equal to one another for $x = a$, we obtain two equations for the unknowns E_1 and ξ_1. Solving these equations and retaining only the terms of lowest order in the parameter D in the solution, we finally find

$$\xi_1 = -\frac{\pi(1 + i)\sqrt{l}}{\sqrt{2\beta k_0}\, a}\left(m_1 + \frac{1}{2}\right). \tag{8.11}$$

A similar formula can be found for ξ_2. Substituting into (8.6) and (8.3), we find the imaginary part of the system:

$$\operatorname{Im}\omega = \frac{\sqrt{\pi}}{8} \cdot \frac{v}{\sqrt{\beta l}} D^{3/2}\left[\left(m_1 + \frac{1}{2}\right)^2 + \left(m_2 + \frac{1}{2}\right)^2\right] \tag{8.12}$$

During the time required for one pass through the resonator the field decreases in the ratio $\exp[-\operatorname{Im}\omega\,(l/v)]$, and the intensity decreases in the ratio $\exp[-2\operatorname{Im}\omega(l/v)]$; here it follows from the smallness of the parameter D that the diffraction losses are small (these losses appear in the exponent) compared with unity. Making use of this, we find the relative magnitude of the diffraction intensity losses during one pass of the light through the resonator:

$$\frac{\sqrt{\pi}}{4}\cdot\frac{1}{\sqrt{\beta}}D^{3/2}\left[\left(m_1+\frac{1}{2}\right)^2+\left(m_2+\frac{1}{2}\right)^2\right],\quad (\beta\sim 1). \tag{8.13}$$

Substituting (8.11) into the solution (8.2), we arrive at the conclusion that, unlike the case of a closed resonator, the field is nonvanishing on the boundary of the active portion of the cross section:

$$\frac{E|_{x=a}}{E_0}\sim\sqrt{D}\left(m_1+\frac{1}{2}\right). \tag{8.14}$$

The field on the boundary of the cross section turns out to be small due to the smallness of the parameter D, and therefore the field distribution over the cross section is little different from the distribution in the case of a closed resonator. The fundamental difference, however, is the existence of diffraction losses, which increase rapidly with an increase of the indices of the transverse modes. This means that there is a sharp predominance of transverse modes having small indices, and their angular divergence is comparable to the diffraction limit.

The spectrum of natural frequencies under conditions (8.8) and (8.9) is similarly practically the same as the spectrum in the case of a closed resonator. However, unlike this case, the natural frequencies are slightly complex. The field damping connected with this leads, as is well known, to a spectral-line spreading equal to 2Im ω. The condition for the existence of transverse modes is the smallness of this spreading compared with the spectral interval between neighboring transverse modes. This condition is satisfied for transverse modes having sufficiently small indices which satisfy the inequality (8.9), and is not satisfied for transverse modes whose indices reach the value† $D^{-1/2}$.

A clear physical interpretation can be given of what has been said above. As is evident from Eq. (8.14), for the condition

† In the case of continuous-wave operation of a practical laser the electromagnetic field in the resonator does not decay, and therefore the frequency is real and the diffraction losses do not lead to spectral broadening. However, as will be shown in Chap. IV, the maximum value of $m_\perp$ does not reach the value $D^{-1/2}$ in this case either.

$m_\perp \ll D^{-1/2}$ the field on the boundary of the active portion of the cross section is small (i.e., the reflectivity of this boundary is close to unity); this is the condition for the existence of a standing wave. However, if $m_\perp$ exceeds the value $D^{-1/2}$, then the field at the boundary is not small, the reflectivity is small, and standing waves cannot develop.

The diffraction losses similarly have a rather clear physical meaning. From a rough estimate it would follow that the magnitude of the diffraction losses, referred to a single pass of the light through the resonator, is approximately equal to D. In fact, during one pass a ray undergoes a transverse shift Δx which is equal to the diffraction angle λ/a multiplied by l ; the fraction of energy which has exited from the resonator during this process is equal to $\Delta x/a = \lambda l/a^2 = D$. Such an estimate turns out to be excessively high for the reason that the nonuniform field distribution over the cross section is not taken into consideration, and, in particular, the smallness of the field near the edges of the cross section is neglected. Thus, the nonuniform field distribution of natural oscillations over the cross section leads to a substantial reduction of the diffraction losses. In particular, the fundamental transverse mode ($m_1 = m_2 = 0$) represents the field distribution which is most advantageous from the point of view of the smallness of the diffraction losses.

In order to clarify what has been said above, we examine the iteration method of finding transverse modes, which was carried out by Fox and Li [86]. According to this method, we begin by postulating a uniform field distribution on the first mirror (Fig. 16, curve 1). Then the Huygens–Fresnel principle is used to find the field which develops on the second mirror after a time interval l/c has elapsed (curve 2), etc., until after a sufficiently large number of iterations the fundamental transverse mode has been formed. The formation of this mode takes place as follows. For the initial field distribution the diffraction losses are relatively high (of the order of D), and they occur basically in that region of the cross section

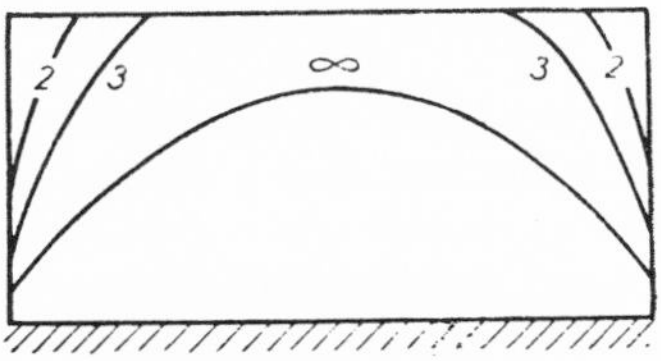

Fig. 16. Graphical representation of the change of the transverse field distribution with an increase in its number of passes between mirrors.

which is adjacent to the cross-section edges. Therefore, during the process of the iterations the field initially decreases more rapidly near the edges than in the central region of the cross section, and the field distribution function undergoes changes which are shown graphically in Fig. 16. Finally, the field on the boundary of the cross section becomes so small that the diffraction losses near the edges of the cross section decrease substantially and become comparable with the losses at the center of the cross section. Therefore, further iterations will no longer change the field distribution. It is precisely this steady-state distribution which is the fundamental transverse mode and has the minimal diffraction losses.

In this chapter we have considered resonators with plane-parallel reflectors. However, in the case of convex reflectors the diffraction losses are practically zero (see Chap. VI for greater detail) for a certain number of transverse modes.

Conclusions

1. In the case of an ideal closed resonator the electromagnetic field on the boundary of the cross section is zero, and there are no diffraction losses even for transverse modes having large indices (the electromagnetic field of these modes propagates at a large angle with respect to the optic axis of the laser).

2. For an open resonator with plane-parallel mirrors the field on the boundary of the active cross section and the diffraction losses are not zero. The diffraction losses increase rapidly with an increasing index of the transverse mode (i.e., with an increase of the angular spread of the light); for a stipulated mode they are proportional to the quantities $(\lambda l/a^2)^{3/2}$.

3. The dependence of the diffraction losses on the transverse mode index leads to the selection of transverse modes having small indices.

4. In the case of a resonator with concave mirrors diffraction losses are practically absent for a certain number of transverse modes, and no selection occurs.

Chapter IV

THE TRANSVERSE STRUCTURE OF THE ELECTROMAGNETIC FIELD IN A PLANE-PARALLEL RESONATOR

In the present chapter we shall consider a practical optical resonator containing an active medium.† The radiation in such a resonator undergoes losses connected with incomplete reflection from the end mirrors and absorption in the host material. These losses, which can be described by the inactive absorption coefficient $\varkappa_1$, are offset by the pump energy. Unlike the previous chapter, we shall not assume that the coefficient $\varkappa_1$ and the pump power N are zero. We shall show that the existence of radiation losses and their compensation by the pumping lead to new phenomena which are absent in the model of an ideal optical cavity resonator. For example, the wave equation describing the electromagnetic field in a practical resonator is nonlinear, but the nonlinear wave equation is satisfied by a definite superposition of longitudinal and transverse modes, rather than by an individual mode or an arbitrary linear combination of modes (as was the case in the previous chapter); this superposition can include a large number of modes. This effect, which is conveniently called nonlinear interaction of modes, determines the most important characteristics of the laser radiation: the spectral composition, the angular divergence, the diffraction losses, and the generation kinetics.

From what follows it will be clear that under certain conditions the two systems of modes (transverse modes and longitudinal modes) interact only negligibly with one another, so that it is sufficient to limit the investigation to the nonlinear interaction within the confines of each of these systems separately. In this chapter we con-

† Schawlow and Townes [101] first suggested that such a resonator could be used to make a laser.

sider nonlinear interaction of transverse modes, while in the next chapter we account of the nonlinear interaction of longitudinal modes.

In order to exclude longitudinal modes from our analysis, we assume that the active medium is homogeneous along the optic axis. [If we assume that the pump distribution is uniform in the longitudinal direction, then our assumption is satisfied, for example, in the case of an active medium which moves parallel to the optic axis with a sufficient velocity [43] (Chap. V)]. Making use of the homogeneity of the active medium along the optic axis, we conceive of an active medium which extends infinitely in this direction. Mathematically, this means that we can ignore the boundary conditions on the end mirrors and can describe the electromagnetic field in the resonator by means of waves traveling along the z axis.

§9. The Equations for the Electromagnetic Field

Haken and Sauermann [88] examined the modes of an optical cavity taking the active medium and the pumping into account. They made the simplifying assumption that the excited atoms are uniformly distributed over the cavity volume. This assumption means that the cavity with the active medium is practically the same as an ideal resonator; however, as shall be shown below, there are a number of fundamental differences.

The point is that it is actually the distribution of the pump intensity over the volume of the active medium which is stipulated, rather than the distribution of the excited atoms. But the distribution of the excited atoms is established in such a way that their concentration is greater near the edges of the resonator cross section than in the interior region, since near the edges the radiation density is less, thus causing a reduction in its emitting action. Because of this the transverse modes having large indices, (which have a considerable intensity near the edges) are amplified to a greater degree than modes having small indices (whose intensity is concentrated basically in the internal region of the cross section). As a result, a large number of transverse modes are generated, which is impossible both in the case of an ideal resonator and in the case of a resonator having a uniform distribution of excited atoms. From the above it is clear that in the equations for the electromagnetic field the gain must depend on the field intensity. The wave equation for an inhomogeneous medium has the form [36]

$$\Delta\vec{E} + \frac{\varepsilon\omega^2}{c^2}\vec{E} - \operatorname{grad}\operatorname{div}\vec{E} = 0. \qquad (9.1)$$

Here the light frequency must be assumed real, since we are treating a continuous process. The quantity ε designates the complex dielectric constant, which depends on the effective absorption coefficient $\varkappa \equiv \varkappa(\omega_0)$:

$$\varepsilon = \varepsilon_0\left(1 + \frac{i\varkappa}{k_0}\right), \quad k_0 = \frac{\omega\sqrt{\varepsilon_0}}{c} \equiv \frac{\omega}{v} \qquad (9.2)$$

(ε_0 is the conventional real dielectric constant). For simplicity, we shall assume that the active medium is optically isotropic, so that ε does not depend on the direction of light polarization.

As was shown in § 5, when diffraction losses are neglected the threshold condition has the form $\varkappa(\omega_0) = 0$; in this case ε would be real. Here we shall consider diffraction losses, and in order for them to be offset the effective absorption coefficient must be negative, while ε must be complex.†

It is necessary to substitute the dependence of $\varkappa$ on field intensity into Eq. (9.2). The specific form of this dependence differs for different level diagrams. In order to avoid carrying out parallel calculations for three- and four-level diagrams, we introduce a convenient substitution which will take account of the difference between these diagrams:

$$\zeta = \begin{cases} \dfrac{N - N^*}{N^*} & \text{for a four-level diagram} \\[2ex] \dfrac{N - N^*}{N^*}\left(1 + \dfrac{\mu_0}{\varkappa_0 + \dfrac{1-r}{l_0}}\right) & \text{for a three-level diagram} \end{cases} \qquad (9.3)$$

† In Eq. (9.1) we neglect the dependence of $\varkappa$ on frequency by making use of the smallness of the spectral width $\delta\omega$ of the laser radiation compared with the half-width $\Delta\omega$ of the luminescence band. If this dependence is considered, then the quantity $\varkappa$ varies by an amount of the order of $\varkappa_1(\delta\omega/\Delta\omega)^2$ within a spectral line; however, the characteristic value of the active absorption coefficient, as we will see below, is equal to $\varkappa_1\zeta$, where ζ is determined according to Eq. (9.3). Therefore, the treatment is applicable for the condition

$$\zeta \gg (\delta\omega/\Delta\omega)^2.$$

Usually, the right side of this inequality is very small, so that this constraint on the pump power is not essential.

Here N is the pump power; N* is its threshold value when diffraction losses are neglected; μ_0 is the absorption coefficient of the unexcited material at the maximum of the band corresponding to the working transition (for a three-level laser) of the impurity; $\varkappa_0$ is the absorption coefficient of the host material; l_0 is the length of the active element; r is the arithmetic mean of the reflectivities of the end mirrors.

Further on we shall examine only the four-level diagram, since all of the results can be generalized automatically for the case of a three-level laser using the substitutions (9.3).

For a four-level laser we have the following result according to Eq. (4.3):

$$\varkappa = \varkappa_1 - Bn = \varkappa_1 - \frac{\varkappa_1}{n^*} n. \tag{9.4}$$

Here B is a constant which can be found from the condition requiring that $\varkappa = 0$ for $n = n^*$ (n^* is the threshold number of excited atoms per unit volume in the absence of diffraction). In order to express $\varkappa$ in terms of the field intensity J, note that for continuous operation

$$\frac{\partial n}{\partial t} = 0 = -\frac{n}{T} - \frac{v\varkappa_1 nJ}{\hbar\omega_0 n^*} + \frac{N}{\hbar\omega_0}. \tag{9.5}$$

(The first and second terms in the right side describe spontaneous and stimulated emission from excited atoms, while the last term describes the excitation of atoms by the pumping.) Determining the value of n from this and substituting it into (9.4), we find

$$\varkappa = -\varkappa_1 \frac{1 - \dfrac{v\varkappa_1 J}{N^*\zeta}}{\dfrac{1}{\zeta} + \dfrac{v\varkappa_1 J}{N^*\zeta}} = -\varkappa_1 \frac{1 - |E|^2}{\dfrac{1}{\zeta} + |E|^2}. \tag{9.6}$$

This expression was simplified by choosing the appropriate unit of measurement for the electric field E.

Substituting (9.2) and (9.6) into Eq. (9.1), we obtain the final equation for the electromagnetic field in a resonator containing an active medium:

$$\Delta\vec{E} + k^2\vec{E}\left(1 - \frac{i\varkappa_1}{k} \cdot \frac{1 - |\vec{E}|^2}{\dfrac{1}{\zeta} + |\vec{E}|^2}\right) - \operatorname{grad}\operatorname{div}\vec{E} = 0, \quad k = \frac{\omega}{v}. \tag{9.7}$$

Let us now examine the boundary conditions for the field.

In the case of an ideal open resonator, as was shown in the previous chapter, the field E_g at the boundary of the active region of the cross section is a small quantity, and it is precisely this quantity which causes the diffraction radiation losses. However, the field distribution over the cross section is practically independent of the small quantity E_g and can be calculated for $E_g = 0$.

It can similarly be shown that in the case of a practical resonator the field at the boundary of the active region of the cross section† is $E_g \sim \sqrt{\xi \varkappa_1 l}$. Hereafter we shall assume that ξ is a small parameter; therefore, it can be assumed that

$$E_g = 0. \tag{9.8}$$

The field distribution calculated for this approximate boundary condition is practically the same as for the true field.

By analogy with the case of an ideal resonator it would seem that an objection could be raised against the boundary condition (9.8) on the basis that it does not allow the diffraction losses to be found, since in a closed resonator in which $E_g = 0$ there are no such losses. In reality, however, this objection is invalidated for a practical resonator. As will be shown below (§ 12), the diffraction radiation losses in a practical resonator are expressed in integral form via the field distribution function, and the small error which the use of the approximate boundary condition (9.8) causes in this function has practically no effect on the magnitude of the diffraction losses.

We have considered the boundary condition at the edge of the resonator cross section. As far as the boundary conditions on the surface of the end mirrors are concerned, we will ignore them in accordance with the statement of the problem in the present

---

† In the case of a practical open resonator (for example, a resonator of the type shown in Figs. 14a and 14b), as well as in the case of an ideal resonator, the field decays outside the active region at a distance of the order of $\sqrt{l/k}$; therefore, the normal derivative of the field is a quantity of the order of $E_g\sqrt{k/l}$. However, in the active region, in which the characteristic value of the field is assumed to be unity, the normal derivative is approximately equal to $1/\delta$ (δ is the characteristic distance). Setting these derivatives equal to one another at the boundary of the active region, we find $E_g \sim \sqrt{l}/\delta\sqrt{k}$. Further on we shall show that for $\xi \ll 1$, $\delta \sim 1/\sqrt{\varkappa_1 k \xi}$, thus, $E_g \sim \sqrt{\xi \varkappa_1 l}$.

chapter. This allows us to seek the solution of the wave equation in the form of a traveling wave.

The boundary conditions on the surfaces of the reflectors will be considered in the next chapter.

In Eq. (9.6) we have chosen a definite normalization of the electromagnetic field in such a way that in the case of an infinitely large cross section and uniform pumping the condition $|E| = 1$ must be satisfied, since for an infinite cross section there are no diffraction losses, and $\varkappa = 0$. It is clear that in the case of a sufficiently large cross section the condition $|E| = 1$ applies everywhere, with the exception of the region next to the cross-section edges, where the diffraction phenomena play a substantial role. Thus, the problem reduces to an investigation of the fields near the edges of the resonator cross section. Therefore, let us begin by examining the case in which the resonator cross section is an infinite half-plane bounded by the straight line $x = 0$; here we assume that the distribution of the pump intensity is uniform [62]. Below we shall show that the case of a laser having a finite cross section and an arbitrary pump distribution over the cross section reduces to this case (it is assumed that the pump intensity is distributed uniformly along the laser axis).

§ 10. Field Distribution over a Cross Section Having the Shape of a Half-Plane

We draw the x and y coordinate axes in the plane of the cross section in such a way that the y axis coincides with the boundary of the cross section, while the x axis is perpendicular to this boundary. The cross section is bounded by the straight line $x = 0$.

It is not difficult to show that for the boundary condition (9.8) the light must be polarized along the boundary of the cross section (i.e., parallel to the y axis). We use E to designate the modulus of the electric field vector of this polarization component. Substituting its field into the wave equation (9.7) and projecting it onto the coordinate axes, we obtain the following equations for E:

$$\frac{\partial^2 E}{\partial x^2} + \frac{\partial^2 E}{\partial z^2} + k^2 E\left(1 - \frac{i\varkappa_1}{k}\cdot\frac{1 - |E|^2}{\frac{1}{\zeta} + |E|^2}\right) = 0, \quad k = \frac{\omega}{v}, \tag{10.1}$$

$$\frac{\partial^2 E}{\partial y \partial z} = \frac{\partial^2 E}{\partial y \partial x} = 0. \tag{10.2}$$

From the last two equations it follows that E does not depend on y.

Far from the boundary of the cross section, as we have already said, $|E| = 1$, and Eq. (10.1) takes the form $E_{xx} + E_{zz} + k_2 E = 0$. It is satisfied by the plane wave $E = \exp[i(k_1 x + k_3 z)]$. Taking this into account, we shall seek the solution of Eq. (10.1) in the form

$$E(x, z) = F(x)\, e^{i(k_3 z + k_1 x)}, \qquad k_1^2 + k_3^2 = k^2, \tag{10.3}$$

where the function F satisfies the following condition at infinity:

$$\lim_{x \to \infty} F(x) = 1 \tag{10.4}$$

In order to simplify Eq. (10.1), we introduce the dimensionless coordinate

$$u = x\sqrt{\varkappa_1 k \zeta}, \qquad x = \frac{u}{\sqrt{\varkappa_1 k \zeta}} \tag{10.5}$$

and assume ζ to be a small parameter:

$$\zeta \ll 1. \tag{10.6}$$

This inequality allows us to neglect the quantity $|E|^2$ compared with $1/\zeta$. Finally, the wave equation (10.1) takes the form

$$iF''(u) - 2qF'(u) + \{1 - |F(u)|^2\} F(u) = 0. \tag{10.7}$$

Here we have introduced the substitution

$$q = \frac{k_1}{\sqrt{\varkappa_1 k \zeta}} = \frac{\sqrt{k^2 - k_3^2}}{\sqrt{\varkappa_1 k \zeta}}. \tag{10.8}$$

Equation (10.7) must be solved for the boundary condition

$$F(0) = 0, \tag{10.9}$$

which is equivalent to the condition (9.8), and for the condition at infinity (10.4). It can be shown that these conditions can be satisfied simultaneously only for a definite value of q equal to $q_0 = 0.686$. For this value of q Eq. (10.7) with the boundary conditions (10.9) and (10.4) determines a universal function F(u), which can be tabulated. The graph of this function is shown in Fig. 17.

From the figure it is evident that the field intensity, which is zero at the boundary of the cross section, increases monotonically with distance from it and in practice reaches its maximum value

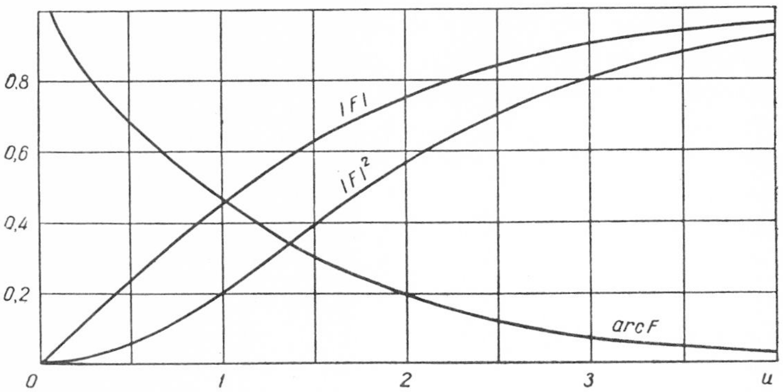

Fig. 17. Transverse distribution of the electromagnetic field near the boundary of the active region of the cross section (u is the distance of the point from the boundary, expressed in $1/\sqrt{\varkappa_1 k\zeta}$ units).

(unity) for $u > 3$ (i.e., for $x > 3/\sqrt{\varkappa_1 k\zeta}$). Thus, the diffraction phenomena affect the intensity distribution of the directional radiation only over a band having a width

$$\delta = \frac{3}{\sqrt{\varkappa_1 k\zeta}}, \tag{10.10}$$

which is adjacent to the boundary of the cross section. Outside this band the diffraction phenomena are insignificant, and therefore the intensity distribution is determined by the laws of geometric optics. In fact, for $x > \delta$ the value of the intensity, which is practically equal to unity, is not altered if we place $\lambda = 0$.

The physical meaning of the results is that, unlike the case of an ideal resonator, the field distribution in the active medium is determined not only by the diffraction phenomena but also by the pumping. On the one hand, the diffraction tends to reduce the intensity of the generated radiation near the edges of the resonator cross section, while on the other hand the distribution of this intensity tends to approach the distribution of the pump intensity.

We shall trace the mechanism of this competition in greater detail. If there were to be no diffraction losses, then the entire excess pump power $N - N^*$ would be converted into generated radiation, and the distribution of its intensity J would be uniform for uniform pumping: $J = J_0 = \text{const}$. The diffraction losses lead to a reduction in intensity by the amount ΔJ, which must be determined from the condition for compensation of the diffraction losses at the expense of the pumping. If the intensity decreases by the amount ΔJ, then the power required to compensate the conventional (nondiffraction) light losses will be reduced correspondingly (i.e., in the ratio $(J_0 - \Delta J)/J_0$), and the incremental threshold above-pump power, which is equal to $(N - N^*)\Delta J/J_0$, will be expended to compensate for the diffraction losses. From this it follows that the reduction of the intensity ΔJ is proportional to the magnitude of the diffraction losses and inversely proportional to the pump power above-threshold $N - N^* \equiv N^*\zeta$. Away from the boundary of the cross section, the diffraction losses decrease,† and $\Delta J \to 0$; therefore, diffraction has practically no effect on the intensity distribution of the radiation far from the edges. Near the boundary of the cross section the diffraction losses are so great that almost the entire threshold power is required to offset them; therefore $\Delta J \sim J_0$ (i.e., the diffraction leads to a substantial reduction of the intensity). The width of the region in which the diffraction phenomena are substantial decreases with decreasing wavelength and with increasing pump power above-threshold $N - N^* = N^*\zeta$. This is evident from Eq. (10.10), if we recall that the threshold pump power is proportional to $\varkappa_1$ in the absence of diffraction losses N^* (see § 5).

§ 11. The Intensity Distribution over a Cross Section Having Finite Dimensions

It was shown above that the distribution of the electromagnetic field over the resonator cross section is established as a result of competition between the effects produced by diffraction and by the bahavior of the active medium excited by the pumping. This competition determines the width δ of the band within which diffraction phenomena are substantial. If $\delta \gg a$, where a is the size of the cross section, then the role of diffraction turns out to be predominant over the entire cross section; in other words, the effect of the active medium can be neglected, and the distribution of the electromagnetic field is practically no different from the distribution in the case of an ideal resonator (Chapter III). Conversely, for $\delta \ll a$ the effect of the active medium turns out to be predominant, and the field distribution differs substantially from the distribution in the case of an ideal resonator.

† Unlike the case of an ideal resonator, we can speak here of local diffraction losses (see § 12).

A clear interpretation can be given to what has been said above. The electromagnetic field of the fundamental transverse mode propagates at an angle of $\theta \sim \lambda/a$ with respect to the optic axis. In order to cross the resonator in the transverse direction the light must traverse a distance of the order of $a/\theta \sim a^2/\lambda \sim ka^2$ through the material in the direction of the optic axis and be amplified in the ratio $\exp(-a^2k\varkappa) \sim \exp(a^2k\varkappa_1\zeta) \sim \exp(a^2/\delta^2)$. For $\delta >> a$ the light is amplified insignificantly in traversing the cross section of the resonator, and therefore the field distribution of the natural oscillations over the cross section is practically no different from the distribution in the case of an ideal resonator in which there is no amplification (or absorption). However, if $\delta \ll a$, then the amplification of the light crossing the resonator cross section substantially changes the transverse field distribution compared to the distribution in the case of an ideal resonator.

Let us examine these limiting cases in greater detail.

1. For the condition $\delta \gg a$, as we have already said, the transverse natural oscillations are close to the transverse modes of an ideal resonator.† Because of the increase in diffraction losses with an increase in the indices of the transverse modes, oscillation can be sustained in only the fundamental or at the most in the modes close to it.

Small corrections to the field distribution and the frequencies of the natural oscillations, which were connected with the effect produced by the active medium, were obtained in [76].

Transverse modes with small indices were observed experimentally in a ruby laser [30, 125].

2. The second case we will consider is

$$\frac{\delta}{a} = \frac{3}{a\sqrt{\varkappa_1 k\zeta}} \ll 1 \qquad (11.1)$$

Since bands having a width δ in which diffraction phenomena are substantial have a spacing which considerably exceeds their width, it follows that the interaction of these regions can be neglected;

† The difference from an ideal resonator resides merely in the fact that for continuous-wave operation the frequency is real, so that the diffraction losses do not lead to spectral broadening. Note that the transverse modes are actually close to the ideal ones even for $\delta \sim \alpha$ [108].

therefore, the distance between bands is physically infinite, and the field distribution near the boundary of the cross section is stipulated by the equations of the preceding section. The curvature of the line which serves as the boundary of the cross section does not impede the applicability of these results, since in accordance with the inequality (11.1) the radius of curvature considerably exceeds δ (if the boundary is a broken line, then the treatment is inapplicable near the vertices).

The treatment can be generalized for the case of a nonuniform distribution of the pumping, provided that the natural assumption is made to the effect that the characteristic dimension of this nonuniformity coincides with the dimension a of the cross section. Then it follows from the condition (11.1) that within the limits of the band having the width δ in which diffraction phenomena are substantial the pump intensity distribution can be treated as uniform. Thus, near the edge of the cross section the results obtained above remain in force. But in the inner region of the cross section, whose points are further than δ from the boundary, diffraction phenomena play practically no role, and the intensity distribution of the stimulated emission replicates the distribution of the pump intensity above-threshold. Actually, the pumping must offset only those losses that occur in addition to diffraction losses in the inner region; these losses are connected with light absorption in the material and incomplete light reflection from the mirrors (i.e., $N - N^* = v\varkappa_1 J(\vec{r})$, where $J(\vec{r})$ is the volume energy density of the stimulated emission; N is the pump power absorbed per unit volume; $\vec{r}$ is a point of the cross section). (For simplicity, we assume that the energy luminescence yield is equal to unity.) From this it follows that in the inner region of the cross section we have

$$J(\vec{r}) = \frac{N(\vec{r}) - N^*}{v\varkappa_1} \quad \text{for} \quad x \gg \delta \tag{11.2}$$

(x is the distance of the point $\vec{r}$ of the cross section from its nearest boundary).

Summarizing what has been said above, we obtain the equations for the volume energy density of stimulated directional emission in the resonator [62]:

$$J(\vec{r}) = \frac{N(\vec{r}) - N^*}{v\varkappa_1} \left| F\left(x\sqrt{\varkappa_1 k\zeta(\vec{r})}\right)\right|^2 \tag{11.3}$$

(the function $|F(u)|^2$ is shown in Fig. 17). If the distance x of the point from the nearest boundary of the cross section exceeds δ, then this equation becomes Eq. (11.2), written without considering diffraction (i.e., in the geometric-optics approximation ($\lambda \to 0$)); however, if $x \lessdot \delta$, then Eq. (11.3) coincides with the results of the preceding section, which were obtained for a narrow band adjacent to the edge of the cross section. Thus, the first factor in Eq. (11.3) represents the intensity of the stimulated emission in the limit of geometric optics, while the second factor describes the decrease in intensity due to diffraction. As the distance from the boundary of the cross section increases, the intensity increases from zero to the value corresponding to the limit of geometric optics. In particular, for uniform pumping the intensity distribution is uniform at a distance $x > \delta$ from the edge of the cross section.

This result constitutes the fundamental difference from the case of an ideal resonator, in which the intensity distribution is essentially nonuniform over the entire cross section. This difference is connected with the fact that the intensity distribution in an ideal resonator is determined exclusively by diffraction, while in a practical resonator it is determined by the competition between the effects of diffraction and pumping (here the effect of pumping is decisive for condition (11.1)). As we have already stated, this competition is manifested in the fact that δ decreases with an increase of the pumping and a decrease of λ.

The above can be interpreted in terms of transverse modes. In the case of an ideal resonator the wave equation is linear, and its solution is any of an infinite number of modes or an arbitrary linear combination of them. In the case of a practical resonator the field is described by a nonlinear equation whose solution is unique for stipulated boundary conditions; it represents a completely defined (but not an arbitrary) superposition of transverse modes of an ideal resonator in which transverse modes having large indices $m_\perp$ play an essential role. The characteristic value of the index $m_\perp$ can be estimated from those concepts according to which the characteristic distance $a/m_\perp$ for the δ-th mode must have an order of magnitude which coincides with δ. From this we find the characteristic value of the index of the transverse mode:

$$m_\perp \sim a \sqrt{\varkappa_1 k \zeta}. \qquad (11.4)$$

The physical cause of the generation of a large number of transverse modes, regardless of the diffraction losses which increase rapidly with the index of the mode, resides in the fact that modes with large $m_\perp$ have a considerable intensity in the region adjacent to the boundary in which the decay of the total intensity of the generated light leads to a large gain.†

As we have already mentioned, the angular divergence of the radiation exceeds the diffraction angle $m_\perp$ by a factor of λ/a; from this it is possible to find the divergence angle of the radiation by means of (11.4): $\theta \sim \sqrt{\varkappa_1 \zeta / k}$. This same result can be obtained differently. Near the boundary of the cross section the angle k_1/k formed by the direction of light propagation with the optic axis has the following form in accordance with Eq. (10.8):

$$\theta = q_0 \sqrt{\frac{\varkappa_1 \zeta}{k}} = 0.7 \sqrt{\frac{\varkappa_1 \zeta}{k}} . \tag{11.5}$$

This is the characteristic magnitude of the angular divergence.

§ 12. The Diffraction Losses

Let us consider the following limiting cases.

1. Assume $a\sqrt{\varkappa_1 k \zeta} \ll 1$. In this case only the fundamental mode and the transverse modes close to it are generated, and the practical resonator is close to the ideal resonator.

As we have already said in § 8, the diffraction radiation losses in an ideal resonator are defined as the damping of the electromagnetic field in the resonator, which is proportional to Im ω. In the continuous-wave case considered here this definition is inapplicable, since the frequency is real; however, it is possible to speak of the

† The strong effect of the active medium, as expressed by the inequality (1.1), leads not only to a large number of generated transverse modes, but also to the modification of each transverse mode having an index of the order of $a\sqrt{\varkappa_1 k \zeta}$ relative to the case of an ideal resonator. This is not difficult to verify [108] if we examine the natural oscillations of a resonator having a cross section in the shape of an infinite strip with a width a and a somewhat simplified transverse distribution of the effective absorption coefficient: $\varkappa = -\varkappa_1 \zeta$ for $x < \delta$ and $a - x < \delta$; $\varkappa = 0$; for $\delta < x < a - \delta$. From what has been said above it follows that for condition (11.1) it is impossible to use the method of expanding the electromagnetic field in modes of an ideal resonator [4, 5, 105].

relative magnitude of the diffraction losses of light energy during a single pass through the resonator. For this value Eq. (8.13) remains in force.

The quantity (8.13) depends on the indices of the transverse mode. If many transverse modes are generated, then it is more convenient to consider the diffraction losses for the entire radiation field as a whole rather than for each mode separately. Let us examine this case in greater detail [62].

2. Assume $a\sqrt{\varkappa_1 k\zeta} \gg 3$, i.e., a large number of transverse modes are generated. Let us define the diffraction losses as the relative decrease in the stimulated-emission energy due to diffraction:

$$\Gamma = (\mathscr{E}_0 - \mathscr{E})/\mathscr{E}_0. \tag{12.1}$$

Here $\mathscr{E}$ is the total stimulated-emission energy in the resonator; $\mathscr{E}_0$ is the same energy for $\lambda \to 0$ (i.e., when there is no diffraction).

Note that in the case of an ideal resonator the damping is introduced for the field as a whole; therefore, it would be meaningless to speak of local diffraction losses at a given point in the cavity. Instead, the diffraction losses differ for different modes of the ideal resonator. But the definition (12.1) retains its meaning for an individual point on the cross section; as we have already stated, the diffraction losses decrease with distance from the boundary of the cross section. At the same time, this definition applies directly to the entire ensemble of transverse modes which are generated by a practical resonator.

Assuming that condition (11.1) is satisfied, we find the following equation by means of Eq. (11.3) [62]:

$$\Gamma = \frac{b}{\sqrt{\varkappa_1 k}} \cdot \frac{\int [N(\vec{r}) - N^*][\zeta(\vec{r})]^{-1/2} dL}{\int [N(\vec{r}) - N^*] dS}, \tag{12.2}$$

where†

$$b = \int_0^\infty \{1 - |F(u)|^2\} \, du = 2.01 .$$

Integration is carried out over the contour of the cross section in

† In an actual experiment b is somewhat less than 2.01, since the field does not vanish on the boundary of the active region.

the numerator of Eq. (12.1), whereas integration is carried out over its area in the denominator.

For a circular cross section having a radius a and an axially symmetrical distribution of the pumping Eq. (12.2) takes the form

$$\Gamma = \frac{ba}{\sqrt{\varkappa_1 k \zeta(a)}} \cdot \frac{N(a) - N^*}{\int\limits_0^a [N(r) - N^*]\, r dr}. \qquad (12.3)$$

But in the case of uniform pump distribution and a cross section having an arbitrary shape, we have

$$\Gamma = \frac{b}{\sqrt{\varkappa_1 k \zeta}} \cdot \frac{L}{S}, \qquad (12.4)$$

where S is the area of the cross section; L is the perimeter of the cross section.

The physical meaning of the latter equation is quite obvious. The diffraction losses decrease with an increase of the characteristic dimension S/L of the cross section, with a decrease of the light wavelength, and also with an increase of the pump power and the nondiffractional losses (the latter facilitate the reduction of the role played by diffraction phenomena).

From a comparison of Eqs. (11.4) and (12.4) it follows that $m_\perp \sim 1/\Gamma$; i.e., as the diffraction losses decrease, the number of transverse modes involved in generation increases.

§13. The Range of Applicability of the Treatment. Comparison with Experiment

For purposes of simplifying the computations we assume that $\zeta \ll 1$. It is easy to see that for $\zeta > 1$ a further increase in ζ will not lead to substantial changes in the field distribution. Actually, the parameter ζ is included in the wave equation (9.7) in the combination $\zeta^{-1} + |E|^2$, but the characteristic value of the intensity $|E|^2 = |F|^2$ in the decay range ($x < \delta$) is a quantity of the order of unity (approximately 0.3 to 0.5), as is evident from Fig. 17. Therefore, for $\zeta \gg 1$ the final equations obtained above remain in force if we assume that the parameter ζ in them is equal to two or three. Under these conditions we must remember that, in accordance with the definition (9.3), ζ is introduced differently for the three-level and four-level diagrams.

In order to illustrate the above let us present experimental data on the angular divergence and output power of a plane-parallel resonator† with an active element consisting of neodymium–glass (these data were taken from [5]). In the paper indicated the experiments were carried out well above the generation threshold; here, in accordance with what has been said above, the experimental values of the angular divergence θ are practically independent of ξ for $\xi \gg 1$.

Experimental investigations of the angular divergence, which were carried out for a large range of resonator lengths, showed that the angular divergence θ is proportional to $l^{-1/2}$ (i.e., to $\sqrt{\varkappa_1}$) with sufficient accuracy. In Fig. 18a the crosses show the experimental values of θ/θ_D, where $\theta_D = 0.3\ \lambda/a$ is the diffraction angle (a is the radius of the active rod). The straight line in this figure was drawn according to Eq. (11.5), in which, with allowance for the population inversion above the generation threshold, we place $\xi = 2.6$. For comparison purposes the dashes show the results of a calculation based on expanding the fields in modes of the ideal resonator [4, 5]. From the figure it is evident that the experimental data are in good agreement with Eq. (11.5). A certain divergence occurs for large t, for which the criterion (11.1) is poorly satisfied (for $l = 100$ cm the left side of inequality (11.1) is equal to 0.6).

From Eq. (11.5) it follows that θ is independent of a for fixed values of the remaining parameters. An experiment with irising of the active element confirmed this conclusion: for $l = 17.5$ and $a = 0.5$ cm we have $\theta = 98'$, and for the same l and $a = 0.3$ cm we have $\theta = 93'$.

The paper indicated similarly presents experimental data on the angular divergence for a laser with an active element consisting of $CaF_2:Sm^{++}$, which has a large light-scattering coefficient. In this case the experimental values of θ similarly fit the curve $\theta = Cl^{-1/2}$ well, but due to light scattering the coefficient C turns out to be somewhat greater than the value which would be obtained from Eq. (11.5).

---

† Such experiments are substantially complicated by the fact that even the weak effective lens that develops as a result of nonuniform heating of the sample radically alters the angular distribution of the light (see §24).

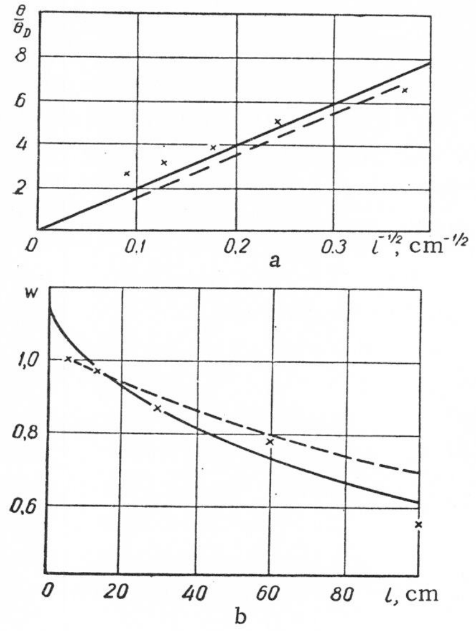

Fig. 18. Comparison of the theoretical transverse structure of the field with experimental data.

Figure 18b shows the experimental data on the output power W for a neodymium–glass laser [5]. The curve $W = \text{const}\,(1 - \Gamma)$ is plotted in accordance with (12.4) for $b = 1.3$. From the figure it is evident that the theory presented completely explains the experimental relationships (for comparison purposes in Fig. 18b dashes show the function calculated by the method of expansion in modes of an ideal resonator [5]).

Conclusions

1. In a resonator containing an active medium the nonlinear interaction of transverse modes is characterized by the ratio a/δ, where $\delta = 3/\sqrt{k\varkappa_1\xi}$. For $\delta \gtrsim a$, the resonator is close to being ideal, and only those transverse modes having small indices (predominantly the fundamental transverse mode) are generated.

2. If $\delta << a$, then the laser differs substantially from a laser with an ideal resonator: a considerable number of transverse modes (a number of the order of a/δ) go over into generation; here the modes having large indices differ from the modes of the ideal resonator. The diffraction phenomena affect the transverse distribution of the integrated intensity only within the limits of the region hav-

ing the width δ, which is adjacent to the boundary of the cross section. Outside this region the transverse distribution of the integrated intensity of the generated light replicates the distribution of the pump intensity. The magnitude of the diffraction losses of the integrated intensity is $\Gamma \sim \delta/a$, and the angular divergence is $\theta \sim \lambda/\delta$.

3. In the case of a three-level diagram the nonlinear interaction of the modes is manifested more strongly, and the number of simultaneously generated modes is larger than it is in the case of a four-level laser having analogous parameters.

Chapter V

LONGITUDINAL STRUCTURE OF THE ELECTROMAGNETIC FIELD IN A PLANE-PARALLEL RESONATOR

In the preceding chapter we considered a practical optical resonator (i.e., a resonator containing an active medium excited by pumping having a given intensity distribution). The electromagnetic field in such a resonator can be described by a nonlinear wave equation whose solution is a completely defined linear combination of modes (or natural oscillations). The effect of the simultaneous generation of a certain number of transverse modes, which is connected with the linear interaction of these modes, was considered in the preceding chapter. There we assumed that the active medium was uniform in the direction of the optic axis and did not consider the boundary conditions on the end mirrors; thus, we ignored longitudinal modes.

In the present chapter we shall investigate the longitudinal structure of the electromagnetic field formed by a system of longitudinal modes whose simultaneous generation is connected with their nonlinear interaction. In order to consider longitudinal modes we shall take account of the boundary conditions for the electromagnetic field on the surface of the end reflectors (the field on this surface must vanish). Then we shall show that the longitudinal and transverse structure of the field in a plane-parallel resonator can be considered independently, since the systems of longitudinal and transverse modes do not interact with each other in practice.†

The nonlinear interaction of longitudinal modes is physically connected with the fact that a high concentration of excited atoms

† This is not applicable to a spherical resonator having a sufficiently large curvature (see §22).

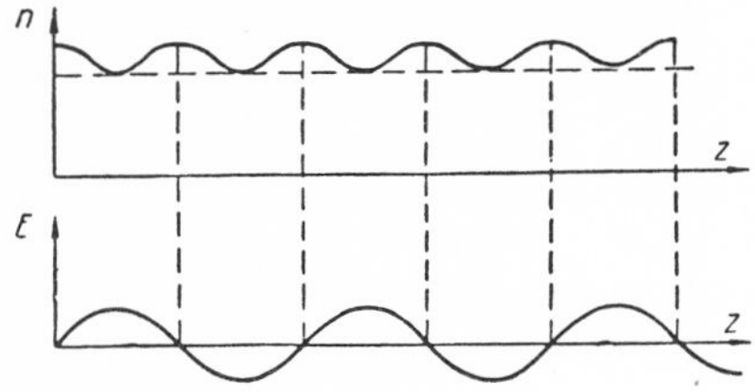

Fig. 19. Nonunifrom spatial distribution of the field intensity (bottom curve), and nonuniform distribution of the population inversion (upper curve).

develops near the nodes of the standing waves (Fig. 19); this makes it possible to generate other standing waves as well, for which the intensity maxima more or less overlap the gain maxima.

Although in general the nonlinear interaction between longitudinal modes is of the same physical nature as the interaction between transverse modes, the effect of the active medium produces practically no modification of the longitudinal modes (unlike the transverse modes) with respect to those of an ideal resonator. This difference is connected with the fact that the longitudinal modes are separated by considerably greater spectral intervals than those which separate the transverse modes. This can be explained more clearly on the basis of the fact that a light ray changes its intensity to a considerably lesser extent during one longitudinal pass through the resonator than during one transverse pass (the latter is accompanied by many longitudinal passes).

We shall make use of this and describe the field distribution along the optic axis in the form of an expansion in longitudinal modes of an ideal resonator (i.e., in sinusoidal standing waves).

§ 14. The Longitudinal Mode of a Resonator Having an Infinite Cross Section

Let us first consider the case in which only one longitudinal mode is generated. The result will later be generalized for the case of many modes. An individual longitudinal mode is also of independent interest, since it can be isolated by various experimental methods (§ 15).

For a resonator with an infinite cross section the field depends solely on the longitudinal coordinate z, and the wave equation (9.7) takes the following form for the condition $\zeta \ll 1$:

$$E''(z) + k_3^2 E - ik_3\varkappa_1\zeta\{1 - |E(z)|^2\}E = 0; \quad k_3 = \frac{\pi m_3}{l}. \tag{14.1}$$

This equation must be solved for the boundary conditions

$$E(0) = E(l) = 0. \tag{14.2}$$

If the gain of the active medium were to be independent of z, then the effective absorption coefficient would be zero, as was shown in §§ 5 and 6 for a homogeneous active medium. Then the last term would not appear in Eq. (14.1), and this equation would be satisfied by an ordinary standing wave $E = \sin k_3 z$. The inhomogeneity of the active medium is considered by the last term in Eq. (14.1); this term is smaller than the previous ones in the ratio $\varkappa_1/k \sim 10^{-5}$. The small term in the equation must correspond to a small correction in the expression for the field; therefore, we shall seek the solution in the form

$$E = E_0 \sin\left(k_3 z - \frac{i\varkappa_1}{k_3} A \sin 2k_3 z\right), \quad k_3 = \frac{\pi m_3}{l}, \tag{14.3}$$

where E_0 and A are constants which must be determined. In this notation we have considered the boundary conditions, as well as the fact that the concentrations of excited atoms are a periodic function of z with a period $\lambda/2$ (see Fig. 19).

After substituting this solution into Eq. (14.1) and carrying out linearization with respect to the small parameter $\varkappa_1/k$, the equation takes the form

$$-4A \sin 3k_3 z + \zeta\left[\left(1 - \frac{3}{4}|E_0|^2\right)\sin k_3 z + \frac{|E_0|^2}{4}\sin 3k_3 z\right] = 0. \tag{14.4}$$

From this we find: $E_0 = 2/\sqrt{3}$, $A = \zeta/12$. Thus, the electromagnetic field of one longitudinal mode has the form

$$E = \frac{2}{\sqrt{3}}\sin\left(k_3 z - \frac{i\varkappa_1\zeta}{12k_3}\sin 2k_3 z\right), \quad k_3 = \frac{\pi m_3}{l}. \tag{14.5}$$

In accordance with the nonlinear character of the wave equation the field has a rigorously specified amplitude. We recall that the normalization of the field is chosen in such a way that in the case of a uniform distribution of the field intensity the magnitude of the intensity $|E|^2$ is equal to unity. However, in the case investigated, in which the intensity is distributed nonuniformly, its maximum value is $|E_0|^2 = 4/3 > 1$ and its average value is $2/3 < 1$.

The basic result of the solution is that the amplitude E_0 is specified; the correction to the argument of the sine is very small and can be neglected. At first glance it might seem strange that the field amplitude remains fixed for an arbitrarily small magnitude of the nonlinear term in Eq. (14.1). However, this result has a simple physical meaning. The electromagnetic field, in propagating in a medium having an effective absorption coefficient $\varkappa(z) = \varkappa_1 \zeta (1 - |E|^2)$, increases its energy by the amount $\nu \int \varkappa |E|^2 dz$ in a unit time. Substituting $E = E_0 \sin k_3 z$ into this expression in the zero approximation, we have

$$-\int_0^l \varkappa |E|^2 dz = \varkappa_1 \zeta E_0^2 \int_0^l (1\ E_0^2 \sin^2 k_3 z) \sin^2 k_3 z dz =$$

$$= \frac{\varkappa_1 \zeta E_0^2 l}{2} \left(1 - \frac{3E_0^2}{4}\right).$$

For continuous-wave operation this quantity must be zero for an arbitrarily small value of $\varkappa_1 \zeta$, since in the converse case the field intensity would vary with time; from this we have $E_0^2 = 4/3$.

Thus, in order to find the solution in the zeroth approximation, it is sufficient to require that the energy drawn from the medium by the electromagnetic field must equal zero. In the next section we shall make use of this observation.

§15. The Number of Simultaneously Generated Longitudinal Modes

The problem of the number of simultaneously generated longitudinal modes was considered in [106] for a three-level diagram, as well as in [42–44]. For our analysis we shall use the terms adopted above and shall represent the results in a form which is applicable to any level diagram. For simplicity, we shall assume that $\zeta \ll 1$, having in mind the comments made in § 13; note that, as shown in [44], the treatment is easily generalized for the case of arbitrary values of ζ by substituting ζ for $\zeta/(\zeta + 1)$.

Above it was shown that the average field intensity of a single longitudinal mode is 1.5 times as small as the value of the uniformly distributed intensity; this is connected with the incomplete utilization of the energy of the excited atoms (Fig. 19). This nonuniformity is smoothed by the simultaneous generation of several modes.

For simplicity we shall assume that the frequency of one of the longitudinal modes having the index $m_3 = 2l/\lambda$ coincides with the gain maximum ω_0. We shall call this mode the central mode and shall measure the indices of other modes with respect to it. The central mode shall be labeled with the index 0, the two modes on either side of it shall be labeled with the indices ± 1, etc. The frequency interval between neighboring modes is equal to ω_0/m_3, so that the frequency of the p-th mode is equal to $\omega_p = \omega_0[1 + p/m_3]$, while its wave vector is $k_p = k_0 + \pi(p/l)$ $(p = \ldots -2, -1, 0, 1, 2, \ldots)$.

In order to find the intensity distribution over the mode we shall use the observation made at the end of the preceding section and shall set the energy drawn from the active media to zero for each mode, i.e., we assume that

$$\int_0^l \varkappa(\omega_p)\,|E_p(z)|^2 dz = 0. \quad (15.1)$$

In order to consider the dependence of the effective absorption coefficient $\varkappa$ on frequency, we expand it in a series in powers of $\omega - \omega_0$ making use of Eq. (4.3):

$$\varkappa(\omega) = \varkappa(\omega_0) + [\varkappa_1 - \varkappa(\omega_0)]\frac{(\omega - \omega_0)^2}{\bar{\omega}^2}. \quad (15.2)$$

Here $\bar{\omega}$ is the quantity of the order of the half-width of the luminescence band $\Delta\omega$ (for a Gaussian luminescence band we have $\bar{\omega} = 0.6\,\Delta\omega$). Substituting $\varkappa(\omega_0) = -\varkappa_1\zeta(1 - |E|^2)$, $\omega_p - \omega_0 = p\omega_0/m_3$ into this expression, we have

$$\varkappa(\omega_p) = \varkappa_1\zeta\{-1 + |E|^2 + [1 + (1 - |E|^2)\zeta]\eta p^2\}. \quad (15.3)$$

Here we have introduced the small parameter of the theory:

$$\eta = \frac{\omega_0^2}{\bar{\omega}^2 m_3^2 \zeta} = \frac{1}{\zeta}\left(\frac{\pi v}{l\bar{\omega}}\right)^2. \quad (15.4)$$

The smallness of this parameter derives from the fact that for a solid-state laser the ratio $\omega_0/\bar{\omega}$ does not exceed 1000, while the index of the longitudinal mode is $2l/\lambda \gtrsim 10^5$. Below it will be shown that the maximum value of p is a quantity of the order of $\eta^{-1/3}$; therefore, along with the inequality $\eta \ll 1$ we also have $\eta p^2 \lesssim \eta^{1/3} \ll 1$.

Since $\zeta \ll 1$ by assumption, it follows that in Eq. (15.3) it is possible to neglect terms containing the product of the small parameters η and ζ. Substituting into (15.1), we obtain the final equations for the intensity distribution over the mode:

$$\int_0^l [-1 + |E(z)|^2 + \eta p^2] |E_p(z)|^2 dz = 0 \tag{15.5}$$

B_p designates the field amplitude of the p-th mode; then the field intensity has the following form in the zero approximation in $\varkappa_1/k$:

$$|E_p(z)|^2 = |B_p|^2 \sin^2\left(k_0 + \frac{\pi p}{l}\right) z = \frac{|B_p|^2}{2}\left\{1 - \cos\left(k_0 + \frac{\pi p}{l}\right) 2z\right\}. \tag{15.6}$$

However, the total intensity is $|E|^2 = \Sigma |E|^2$, since the terms in the intensity which correspond to the product of the fields of different modes p and p' have a time multiplier exp it $(\omega_p - \omega_{p'})$ and drop out during the time-averaging process.

We substitute (15.6) into Eq. (15.5) while taking account of the fact that a single cosine or the product of two cosines of different arguments drops out during the integration. Finally, we have

$$-1 + \eta p^2 + \frac{1}{2}\sum |B_r|^2 + \frac{|B_p|^2}{4} = 0. \tag{15.7}$$

We have derived the equations for the amplitudes B_r. In order to simplify the notation, we shall assume that these amplitudes are real, which does not cause any essential change. Obviously, amplitudes of symmetrical modes coincide (i.e., $B_{-p} = B_p$). Assume that a total of 2j + 1 symmetrical longitudinal modes is generated, i.e., $-j \leq p \leq j$. Introducing the substitution

$$\sigma = \sum_{r=-j}^{j} B_r^2 = B_0^2 + 2\sum_{r=1}^{j} B_r^2,$$

we find the following result from Eq. (15.7):

$$\frac{1}{4} B_p^2 = 1 - \frac{\sigma}{2} - \eta p^2. \tag{15.8}$$

Here σ still remains unknown. In order to find this quantity we sum both sides of Eq. (15.8) in the limit from $-j$ to j and make use of the well-known formula $1^2 + \ldots + j^2 = j(j + 1)(2j + 1)/6$. We have

$$\frac{\sigma}{4} = \left(1 - \frac{\sigma}{2}\right)(2j + 1) - \frac{\eta}{3} j(j + 1)(2j + 1).$$

Finding σ from this and substituting the results into (15.8), we obtain the intensity distribution over the modes for the condition that the number of generated modes is equal to 2j + 1:

$$B_p^2 = \frac{4}{4j+3}\left[1 + \frac{2\eta}{3} j(j+1)(2j+1)\right] - 4\eta p^2 \cong$$

$$\cong \frac{1}{j} + 4\eta\left(\frac{j^2}{3} - p^2\right). \tag{15.9}$$

As might have been expected, with an increase of the mode number (i.e., of the spectral interval between its frequency ω_p and the "center" frequency ω_0) the intensity decreases monotonically. Provided only that no particular group of modes (see below) is specially isolated, the number of generated modes is determined from the condition $B_{j+1} = 0$. From the smallest of η it follows that $j \gg 1$; making use of this, we find the number of simultaneously generated modes from Eq. (15.9):

$$2j + 1 = \left(\frac{3}{\eta}\right)^{1/3} = \left(\frac{3\zeta\bar{\omega}^2 m_3^2}{\omega_0^2}\right)^{1/3} = (3\zeta)^{1/3}\left(\frac{l\bar{\omega}}{\pi v}\right)^{2/3}. \tag{15.10}$$

Due to the smallness of the parameter η, this number considerably exceeds unity.

Equation (15.10) has a clear physical meaning. As we have already said, the nonlinear interaction of the modes is connected with the fact that one longitudinal mode (i.e., a standing wave) would leave excited atoms at its nodes. As the excess ζ above the threshold power increases, the number of excited atoms above their threshold concentration increases, and the interaction between modes becomes stronger. On the other hand, modes close to the central mode have predominance because of the large gain; here the difference in the gain for neighboring modes will become more significant as the width $\bar{\omega} \sim \Delta\omega$ of the luminescence band decreases. Therefore, the number of generated modes increases with an increase in ζ and $\bar{\omega}$.

From Eq. (15.10) it follows that the longitudinal modes cover (without completely filling) a spectral interval which is equal to

$$\delta\omega_{\parallel} = 2j\frac{\omega_0}{m_3} = \bar{\omega}\left(\frac{3\zeta\omega_0}{m_3\bar{\omega}}\right)^{1/3} = \bar{\omega}\sqrt{\frac{3\zeta}{2j}} = \left(\frac{3\pi v\bar{\omega}^2\zeta}{l}\right)^{1/3} \tag{15.11}$$

This quantity is considerably less than the half-width of the luminescence band.

As we have already said, a large number of generated modes is connected with the attempt to form a uniform spatial distribution of the intensity. This is confirmed by experiment. It can be shown that if the intensity is distributed according to Eq. (15.9) with respect to the longitudinal modes, then its spatial distribution is uniform with a relative accuracy of the order of $\eta^{2/3} \sim j^{-2}$ everywhere except in a small region adjacent to the mirrors.

So far we have neglected the transfer of excitation energy between impurity atoms, which facilitates equalization of the concentration of excited atoms and thus the reduction of the number of generated modes. Livshits and Tsikunov [42] showed that the diffusion of excitations leads to the reduction of the number of longitudinal modes by a factor of $|1 + (4\pi/\lambda)^2 TD_0]^{1/3}$. Here $D_0 \sim (wR^2/6)$ is the diffusion coefficient of the excitation; R is the mean distance between impurity luminescence centers; w is the probability of excitation transfer between two impurity atoms.† According to the estimate given by Livshits and Tsikunov, the diffusion of excitations does not lead to a substantial reduction of the number of generated modes for a ruby laser.

Experimental data on the total width of the spectral band covered by the generated radiation are in fairly good agreement with the theory [6, 106]. However, the experimentally observed laser radiation spectrum consists not of 2j lines corresponding to all of the longitudinal modes which fall within the band of generated frequencies, but of a considerably smaller number of lines. Such a divergence between the theory presented above and the observed spectrum can be explained by the selection of longitudinal modes in a resonator with external mirrors. This selection is caused by the interference of longitudinal field oscillations on the active-element surfaces and reflector backings which are perpendicular to the optic axis [78, 108-111].

Various experimental methods exist for reducing the spectral width of the generation. In [1-3] this was accomplished by

† The probability of migration of excitations can be calculated according to Eq. (3.6) in which it is necessary to identify the indices S and A and to place $P_1 = P_2$ for a three-level diagram. However, in the case of a four-level laser the subscript 1 in Eq. (3.6) applies to the ground-state level, which is designated by the subscript zero in a four-level diagram (Fig. 7a). In view of the absence of experimental data on $0 \rightarrow 2$ transitions, the estimation of w according to Eq. (3.6) is complicated (a method of indirect estimation is presented in Chap. IX).

using a laser with nonresonant feedback which has a continuous radiation spectrum.

In [46] the active element was moved parallel to the optic axis in order to stress noncentral modes. If the velocity of the active medium exceeds the value $\lambda/T \sim 1$ mm/sec, then during the lifetime of the excitations the medium is able to move through a distance which is greater than one wavelength, so that the spatial nonuniformity of the gain is smoothed substantially. Just as does the diffusion of the excitations, this leads to a reduction of the number of generated modes; however, in the above case a considerably greater effect was achieved [43, 46].

Certain authors [87, 98, 126, 127] have achieved the selection of a small number of longitudinal modes by inserting thin transparent plane-parallel plates into the resonator; the thickness of the plates is equal to an integer or semiinteger number of wavelengths of these modes. However, this method requires the use of highly perfected optical systems.

In [17, 19, 20] a dispersive prism was placed inside a resonator with plane mirrors in order to reduce the spectral width of the laser radiation. Due to the dispersion of the light, the condition of normal incidence of rays on the mirror can be satisfied only at a definite frequency near which generation occurs. The advantage of this method lies in the fact that it provides the possibility of varying the operating frequency of the laser smoothly by rotating the reflector or prism.

The method of producing the spectral width by means of lenses placed in the resonator will be presented in §22.

§16. The Electromagnetic Field in a Resonator Having a Finite Cross Section

In the preceding chapter we investigated the system of transverse modes generated by a resonator containing an active medium; under these conditions we assumed that the active medium was uniform along the optic axis and neglected the boundary conditions for the field on the end reflectors, thus ignoring the longitudinal structure of the electromagnetic field. In this chapter, conversely, we have just considered the system of longitudinal modes in a resonator having an infinite cross section while ignoring the conditions for the field on the boundaries of the cross section (thus ignoring the transverse structure of the electromagnetic field). Let us now examine the electromagnetic field in a resonator with a finite cross section while simultaneously taking account of the boundary conditions for the field on the surfaces of the reflectors and on the boundaries of the resonator cross section. It can be shown that the system of longitudinal modes which forms the longitudinal structure of the field and the system of transverse modes which forms its transverse structure practically do not interact with one another. Therefore, in a resonator having finite dimensions the conclusions of § 15 with respect to the number of generated longitudinal modes and the frequency bands covered by them remain valid regardless of the transverse structure of the field; moreover all of the conclusions of the preceding chapter concerning the number of generated transverse modes, the angular divergence, and the diffraction losses remain valid.

Let us give more detailed consideration to the case in which condition (11.1) is satisfied; this condition means that diffraction phenomena affect the distribution of the field intensity only in a narrow band adjacent to the boundary of the cross section [62]. As was shown in the preceding chapter, it is sufficient to examine a cross section in the form of a half-plane. Let us write the boundary conditions on the boundary of the cross section and on the mirrors simultaneously:

$$E\big|_{x=0} = E\big|_{z=0} = E\big|_{z=l} = 0. \tag{16.1}$$

Thereby we consider both the transverse and longitudinal structures of the field.

Let us first examine the case in which only one longitudinal mode is generated (while the remaining ones are suppressed in some way or other). We shall seek the solution of Eq. (9.7) in the form of a modified equation Eq. (14.5):

$$E = \frac{2}{\sqrt{3}} f(x) \sin\psi, \quad \psi = k_3 z + \frac{i\varkappa_1\zeta}{12k} |f(x)|^2 \sin 2k_3 z, \tag{16.2}$$

where $k_3 = \pi m_3 / l$. Let us substitute this expression into Eq. (9.7) while neglecting, as usual, the quantity $|E|^2$ compared with $1/\zeta$ in the nonlinear term. We have

$$\begin{gathered} f(x)(\sin\psi)''_{zz} + f''(x)\sin\psi + 2f'(x)\cos\psi\cdot\psi'_x + f(x)[\cos\psi\cdot\psi''_x - \\ - \sin\psi\cdot(\psi'_x)^2] + k^2 f(x)\sin\psi - i\varkappa_1 k\zeta f(x)\left(1 - \frac{4}{3}|\sin\psi|^2 |f|^2\right)\sin\psi = 0. \end{gathered} \tag{16.3}$$

From what follows it will be obvious that $|f(x)| < 1$. Taking this into account, it is possible to neglect the third term in the left side in Eq. (16.3), since it is smaller than the second term in the ratio $\varkappa_1/k$; we can also neglect the other terms containing ψ_x and ψ''_{xx}. Note further that for fixed x the function $(2/\sqrt{3})\sin\psi$ coincides with the equations (14.5) in which the substitution $\zeta \to \zeta |f|^2$ has been made. But Eq. (14.5) satisfies Eq. (14.1); making use of this equation, we express $(\sin\psi)''_{zz}$ in terms of $\sin\psi$ and substitute it into (16.3). After simple transformations and reduction by the common factor $\sin\psi$, we have

$$f''(x) + \varkappa_1 k\zeta f(x)\left[q^2 - i\left(1 - |f|^2\right)\right] = 0, \tag{16.4}$$

where

$$q = \frac{\sqrt{k^2 - k_3^2}}{\sqrt{\varkappa_1 k\zeta}}, \quad \text{i.e.,} \quad k \equiv \frac{\omega}{v} = k_3 + \frac{q^2\varkappa_1\zeta}{2} = \frac{\pi m_3}{l} + \frac{q^2\varkappa_1\zeta}{2}. \tag{16.5}$$

Using the substitution of variables $f(x) = \exp(iqu) \cdot F(u)$, $u = x\sqrt{\varkappa_1 k\zeta}$, Eq. (16.4) is reduced to the form (10.7), whose solution we already know, and the graph of it is shown in Fig. 17.

Reasoning further in the same way as we did in § 11, we find the volume density of the light energy in a resonator in which one longitudinal mode is generated:

$$J(\vec{r}) = \frac{4}{3} \cdot \frac{N(\vec{r}) - N^*}{v\varkappa_1} \left| F\left(x\sqrt{\zeta(\vec{r})\varkappa_1 k}\right)\right|^2 \sin^2 k_3 z. \tag{16.6}$$

As in § 14, we have neglected small terms of order $\varkappa_1/k$ in the final results. The factor 4/3 in this equation is connected with the fact that for one longitudinal mode the maximum intensity is 4/3, while the average intensity is 2/3 of the values which could be achieved for uniform intensity distribution along the z axis.

In Eq. (16.6) the same notation has been used as in § 11; namely, $\vec{r}$ is the radius vector of a point on the resonator cross section, and x is the distance of this point from the nearest boundary of the cross section. We shall not discuss the physical meaning of Eq. (16.6), since we would have to repeat what was said in § 11 with slight modifications.

Let us now examine a resonator in which the maximum possible number of longitudinal modes, which is equal to $2j + 1 = (3/\eta)^{1/3}$, is generated. Since in this case the longitudinal intensity distribution is practically uniform, its distribution over the cross section is evidently described by Eq. (11.3).

Now it is clear that for the condition (11.1) the spatial distribution of the intensity does not differ from the distribution in the case of an infinite cross section (§ § 14 and 15) over almost the entire cross section, regardless of the number of generated longitudinal modes. The boundaries of the cross section develop only in a small region having the width $\delta \sim 3/\sqrt{k\varkappa_1\zeta}$ in which the intensity increases monotonically with distance from the boundary and reaches a limit corresponding to the case of an infinite cross section. For condition (11.1) the boundaries of the cross section also cannot affect the intensity distribution over the longitudinal modes; therefore, this distribution is stipulated by Eqs. (15.9) and (15.10), just as in the case of an infinite cross section.

Assuming that condition (11.1) is satisfied, let us examine the angular and spectral distribution of the radiation. As in the case in which the active medium is uniform along the laser axis (§ 11), the angular divergence connected with the simultaneous generation of a considerable number of transverse modes is stipulated by Eq. (11.5). Starting from Eq. (16.5), it is possible to show via analogous reasoning that the nonlinear interaction of transverse modes likewise leads to additional lines around each longitudinal line. The frequency range of the group of modes is equal to

$$\frac{1}{2} q_0^2 v \varkappa_1 \zeta = 0.24 v \varkappa_1 \zeta, \quad (16.7)$$

which amounts to the fraction

$$\frac{1}{2\pi} q_0^2 \varkappa_1 l \zeta = 0.08 \varkappa_1 l \zeta \quad (16.8)$$

of the spectral interval between neighboring longitudinal modes. Usually this quantity is considerably less than unity, so that the spectral lines of the longitudinal modes do not overlap one another.†

† In this chapter we use identical zero boundary conditions on the reflector surfaces and the cross-section boundary. It would seem that this contradicts the fact that the longitudinal and transverse distributions of the field differ and that the diffraction losses are determined by the transverse rather than by the longitudinal distribution. The point is that we consider the difference between the boundary conditions on the mirrors and on the cross-section boundaries implicitly by assuming that the resonator is open and that the angle between the direction of light propagation and the optic axis is small. Mathematically, this is expressed in the smallness of the transverse component of the wave vector $k_1 = q\sqrt{\varkappa_1 k \zeta} \ll k$. Equation (10.7), which leads to the transverse intensity distribution shown in Fig. 17, can be satisfied only for $q = 0.686$; the analogous equation for the longitudinal field distribution does not have the analogous solution, since the parameter q appearing in this equation would be equal to $\sqrt{k/\varkappa_1\zeta}$, which is considerably greater than unity.

For condition (11.1) the equations for the diffraction losses obtained in § 12 similarly remain valid.

Note that the spectral broadening and angular spread, in contrast to the diffraction losses, are not fringing effects, since they do not depend on the dimensions of the cross section and take place as a result of the simultaneous generation of many transverse modes.

Concerning the range of values of ξ in which the investigation is applicable, we can repeat what was said in § 13.

§ 17. Comment on the Effective Absorption Coefficient

In all the computations we consider the losses on the mirrors by adding the term $(1 - r)/l$ to the absorption coefficient. The effective absorption coefficient which is introduced in this manner allows the calculations to be simplified and clarified. However, the same results can also be obtained in principle without using the effective absorption coefficient.

It is immediately clear that consideration of the losses on the mirrors by means of the effective absorption coefficient in the energy relationships is fully adequate; we saw this in the example of calculating the threshold pump power (§§ 5 and 6). This also applies to the kinetic equations considered in Chapter VII, which are essentially energy relationships.

Let us show that writing the wave equation with the effective absorption coefficient is fully justified. For example, let us solve Eq. (14.1) without resorting to the effective absorption coefficient. In order to replace the effective absorption coefficient by the true coefficient in the left side of this equation it is necessary to subtract the quantity $i\varkappa_r kE$ from the left side, where $\varkappa_r = (1 - r)/2l$ (r is the reflectivity of one of the mirrors; the second mirror is assumed to be ideal for simplicity). Simultaneously, it is necessary to write the following boundary conditions:

$$E(0) = i\varkappa_r l, \qquad E(l) = 0 \tag{17.1}$$

(the first of them considers small losses on the mirrors).

If we introduce the substitutions $k' = k_3 - i\varkappa_r$, $z' = z + i\varkappa_r (l/k)$, then the equation acquires its previous form, with the difference that k' and z' have replaced k_3 and z throughout. Its solution, which satisfies the boundary conditions (17.1), differs from the solution (14.5) solely by the additional small term in the argument of the sine, which is equal to $i\varkappa_r (l - z)$. This small monotonically varying term has no significance and can be dropped; however, the basic result, which consists in the determination of the field amplitude, remains unchanged.

The situation is analogous with regard to the wave-equation solution which satisfies conditions (17.1) on the mirrors and the zero boundary conditions on the cross-section boundary simultaneously.

Conclusions

1. The nonlinear interaction of longitudinal modes leads to simultaneous generation of a large number of such modes $(3\zeta\bar{\omega}^2 l^2/\pi^2 v^2)^{1/3}$. The number of simultaneously generated modes (i.e., the measure of their nonlinear interaction) increases with an increase of the relative excess of pump power above threshold, and for an identical excess above threshold it is more pronounced for a three-level diagram than for a four-level diagram; this applies to both longitudinal and transverse modes. In the case of a large number of simultaneously generated longitudinal modes the intensity distribution along the optic axis is practically uniform.

The spectrum of the generated radiation consists of individual lines corresponding to longitudinal modes; each is augmented by transverse modes the number of which is connected with the nonlinear interaction of the transverse modes and is equal to $v\varkappa_1\zeta/4$.

3. The reduction of the number of generated longitudinal modes can be achieved by various experimental methods (for example, by inserting a transparent plate or a dispersive prism into the cavity, and also by using a moving active sample).

4. All of the conclusions of the preceding chapter on the interaction of transverse modes and on the diffraction losses and angular divergence of the radiation remain in force. In particular, for the conditions $a\sqrt{\varkappa_1 k\zeta} \gg 1$ the transverse distribution of the radiation intensity practically replicates the distribution of the pump intensity.

Chapter VI

A LASER WITH CONCAVE REFLECTORS

The three preceding chapters were devoted to a laser having plane-parallel mirrors.

Resonators with concave mirrors, in which the electromagnetic field has qualitatively new properties, have similarly been widely used.

The modes of an ideal resonator with concave spherical mirrors have been covered rather extensively in the literature [23, 80, 85, 86, 131]†. The most complete and consistent theory of this problem can be attributed to Vainshtein [23].

In the presentation given below the general results obtained by Vainshtein are generalized for the case of concave reflectors having an arbitrary shape. Simultaneously, we consider the active medium (i.e., actually, we consider nonlinear interactions of the modes of the ideal resonator) by means of the optomechanical analogy developed below.

§18. The Equivalent Mechanical System

Let us show that the equation for the electromagnetic field in a resonator can be given the form of the equation of motion of a simple mechanical system.

Let us consider the resonator formed by two identical concave mirrors having an arbitrary shape. Assume that the surface of the mirrors is described by the equation $z = u(x, y)$ (the position of the origin and the directions of the axes are shown in Fig. 20). We shall assume that the active medium filling the gap between mirrors is bounded laterally by the surface of a cylinder (not necessarily a circular cylinder).

† A summary of the results can be found in [118].

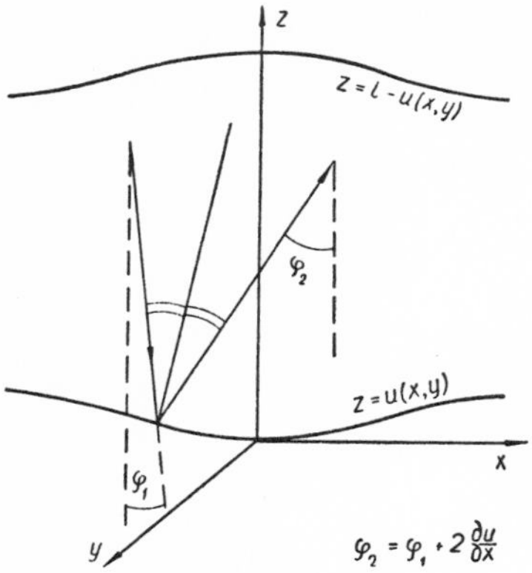

Fig. 20. Diagram for the derivation of the equation of motion of the light in a resonator with concave mirrors in the geometrical optics approximation.

Let us use l to designate the maximum distance between mirrors, while $2a$ designates the diameter of the active sample (or the characteristic dimension of the cross section if it is not round). In order to simplify the calculations, we shall assume that the concavity of the mirrors is small enough so that the condition

$$u \ll \frac{a^2}{l} \tag{18.1}$$

is satisfied.

Let us begin by considering the electromagnetic field in the geometrical optics approximation. Assume $\vec{n}$ is the direction unit vector of a ray, while x, y are the coordinates of the point at which the ray is reflected from the mirror. The ray forms a narrow angle n_x with the yz plane, and this angle changes by the following amount (see Fig. 20) during reflection from the mirror:

$$\Delta n_x = -2\frac{\partial u}{\partial x}. \tag{18.2}$$

From condition (18.1) it follows that a ray must cross the resonator many times in the longitudinal direction before it can be displaced appreciably in the transverse direction. Making use of this fact, we replace the finite differences by derivatives with respect to the number of reflections ν.

We have

$$\frac{dn_x}{d\nu} = -2\frac{\partial u}{\partial x}, \quad \frac{dn_y}{d\nu} = -2\frac{\partial u}{\partial y} \tag{18.3}$$

(the equation for n_y is written by analogy). We also write the obvious equations for the transverse displacement of the ray:

$$\frac{\partial x}{\partial \nu} = l n_x, \qquad \frac{\partial y}{\partial \nu} = l n_y, \tag{18.4}$$

where ν is the number of reflections of the ray from both mirrors.

Differentiating (18.4) with respect to ν, we then eliminate n_x and n_y using Eq. (18.3). Converting to the independent variable $t = l\nu/v$ (t is time and v is the velocity of light in the material), we obtain the equation of motion for the projection of the beam onto the plane of the cross section [63]:

$$\ddot{\vec{r}} = -g \cdot \operatorname{grad} u(x, y), \tag{18.5}$$

$$g = \frac{2v^2}{l} \tag{18.6}$$

($\vec{r}$ is the two-dimensional radius vector of the points (x, y)).

Thus, the motion of the projection of the ray is described by the same equation as that used to describe the motion of a heavy material point along a smooth concave reflector surface (with its concavity facing upward); here the acceleration of gravity must be taken equal to $2v^2/l$.

Obviously, the angular divergence of the radiation is

$$\theta = \frac{V}{v}, \tag{18.7}$$

where V is the velocity of a material point [the inequality (18.1) allows the z component of this velocity to be neglected].

Let us now examine the wave equation for the electromagnetic field using the small parameter $D = \lambda l/a^2$ introduced in § 8. Initially, we shall assume that the active medium is homogeneous (i.e., we shall actually neglect it; the consideration of the active medium and nonlinear interaction of the modes is presented in the next section). Let us write the equation for the scalar electric-field potential in a homogeneous medium:

$$\Delta\Phi(x, y, z) + k^2\Phi(x, y, z) = 0, \qquad k = \frac{\omega}{v}. \tag{18.8}$$

This equation must be solved for the boundary conditions

$$\Phi(x, y, u(x, y)) = \Phi(x, y, l - u(x, y)) = 0, \tag{18.9}$$

which means that the potential vanishes on the surface of the mirror (the mirrors are assumed to be ideal). The conditions at the boundary of the cross section of the active sample will be considered below.

Let us introduce the variable Z instead of z:

$$z=\frac{l}{2}-Z+\frac{2Z}{l}u(\vec{r})\quad\left(-\frac{l}{2}\leqslant Z\leqslant\frac{l}{2}\right). \tag{18.10}$$

On the surface of the reflectors we have $Z = \pm\, l/2$, and the boundary conditions (18.9) take thé simple form

$$\Phi|_{Z=\pm\frac{l}{2}}=0. \tag{18.11}$$

In the nonorthogonal coordinate system x, y, Z the wave equation (18.8) is written as follows:

$$\frac{\partial^2\Phi}{\partial Z^2}\left(1+\frac{4u}{l}\right)+\frac{\partial^2\Phi}{\partial x^2}+\frac{\partial^2\Phi}{\partial y^2}+k^2\Phi+\frac{2Z}{l}\left[2\frac{\partial u}{\partial x}\cdot\frac{\partial^2\Phi}{\partial Z\partial x}+\right.$$

$$\left.+\frac{\partial^2 u}{\partial x^2}\cdot\frac{\partial\Phi}{\partial Z}+2\frac{\partial u}{\partial y}\cdot\frac{\partial^2\Phi}{\partial Z\partial y}+\frac{\partial^2 u}{\partial y^2}\cdot\frac{\partial\Phi}{\partial Z}\right]=0. \tag{18.12}$$

Here we have dropped the terms which are quadratic in the small quantity u and its derivatives. We shall seek the solution in a form which satisfies the boundary conditions (18.11):

$$\Phi(x,y,Z)=\psi(x,y)\sin k_3\left(Z-\frac{l}{2}\right)=\operatorname{Im}\psi e^{ik_3\left(Z-\frac{l}{2}\right)} \tag{18.13}$$

($k_3 = \pi m_3/l$; ψ is a real quantity). Substituting Φ into Eq. (18.12), we have

$$\frac{\partial^2\psi}{\partial x^2}+\frac{\partial^2\psi}{\partial y^2}+k_\perp^2\psi-k_3^2\frac{4u}{l}\psi+\frac{2ik_3Z}{l}\left\{2\frac{\partial\psi}{\partial x}\cdot\frac{\partial u}{\partial x}+\right.$$
$$\left.+\psi\frac{\partial^2 u}{\partial x^2}+2\frac{\partial\psi}{\partial y}\cdot\frac{\partial u}{\partial y}+\psi\frac{\partial^2 u}{\partial y^2}\right\}=0. \tag{18.14}$$

Here $k_\perp$ is the transverse component of the wave vector, i.e.,

$$k\equiv\frac{\omega}{v}=\sqrt{k_3^2+k_\perp^2}=k_3+\frac{k_\perp^2}{2k_3}=\frac{\pi m_3}{l}+\frac{k_\perp^2}{2k}. \tag{18.15}$$

The direction of light propagation forms the small angle

$$\theta\sim\frac{k_\perp}{k}\ll 1 \tag{18.16}$$

with the optic axis.

We shall show that in Eq. (18.14) it is possible to neglect the last term (the term in the braces) in comparison with each of the remaining terms (for example, $k_3^2 \psi u/l$). Noting that $\partial u/\partial x \sim u/a$, $\partial\psi/\partial x \sim k_\perp \psi$, we find that the ratio between the first or second term in the braces and the quantity $k_3^2 \psi u/l \cong k^2 \psi u/l$ is a quantity of the order of $k_\perp l/ka$ or D, respectively. Below it will be shown that $k_\perp^2 \lessapprox k^2 u_{max}/l$. Making use of this, as well as of conditions (18.1), we find that both the first and second of these ratios are considerably less than unity. The remaining terms in the braces are considered similarly.

Thus, the wave equation takes the form†

$$\frac{\partial^2\psi}{\partial x^2} + \frac{\partial^2\psi}{\partial y^2} + \left(k_\perp^2 - \frac{4k^2u}{l}\right)\psi = 0. \tag{18.17}$$

This is none other than the Schrödinger equation for a heavy material point which is located on a smooth concave mirror surface.

Thus, instead of considering the electromagnetic field in the resonator it is possible to consider an equivalent mechanical system: a heavy material point on the reflector surface. The geometrical optics approximation for the electromagnetic field corresponds to the motion of a material point according to Newton's laws of mechanics, and the wave equation for the field corresponds to a quantum-mechanical description of the motion of a material point.

As is well known, a material point can be treated classically if the derivative of its de Broglie wavelength ($2\pi/k_\perp$ in the case given) with respect to the coordinates x, y is small compared with unity (i.e., if the condition

$$k_\perp a \gg 1 \tag{18.18}$$

is satisfied). For the condition (18.18) the motion of the material point can be described by Newtonian mechanics, and the motion of the electromagnetic field in the resonator can be described by the geometrical optics approximation.

The condition (18.18) has a simple meaning: quantum states of the material point (or, correspondingly, transverse resonator

---

† A similar equation was derived by Vainshtein [23] in the particular case of spherical reflectors without the constraint (18.1).

modes) are those that have large indices. If Eq. (18.16) is considered, it becomes clear that the inequality (18.18) also means that the diffraction angle λ/a is small compared with the angular divergence of the radiation. Most often the angular divergence considerably exceeds the diffraction limit, and inequality (18.18) is satisfied. The subsequent treatment is based on this inequality.

§19. Consideration of the Active Medium

Along with the boundary conditions for the field on the surface of the reflectors, it is necessary to examine the conditions on the cylindrical surface (not necessarily having a circular cross section) which serves as the boundary of the active medium. In the case of an open resonator (for example, a resonator of the type shown in Fig. 14b and 14c, except that it has concave mirrors) the conditions at the boundary of the active medium consist in the fact that the electromagnetic field ψ must damp out rapidly beyond the limits of the active medium.

On the other hand, ψ is the wave function of a material point moving in a potential well; it damps out exponentially in the region which is inaccessible to classical analysis and in which the potential energy $4k^2u/l$ exceeds the total energy $k_\perp^2$. For condition (18.18) damping occurs at a distance of the order of $(k_\perp a)^{-2/3}$, which is considerably shorter than the dimension of the cross section. Thus, the condition at the boundary of the cross section takes the form of a constraint on the quantum numbers of the material point (i.e., on the indices of the transverse modes [63]):

$$k_\perp^2 < \frac{4k^2 u_{\max}}{l}, \qquad (19.1)$$

where u_{max} is the value of u on the boundary of the active portion of the cross section (the height of the reflector). The inequality (19.1) is still somewhat schematic; its specific form depends on the shape of the reflectors and the shape of the cross section of the active sample.

Among the modes generated by a laser there must be modes having the maximum value of $k_\perp$ allowed by the inequality (19.1); such modes have a considerable intensity over the entire cross section of the active sample and damp out exponentially outside the cross section. If there were to be no such modes, and only those modes were generated which had smaller values of $k_\perp$ and had a noticeable intensity only in a certain inner region of the cross section, then only this inner region would generate. Outside it the gain would be noticeably greater than inside it, and this would stimulate the generation of modes with large indices, which have a considerable intensity in the region having the large gain. We again encounter the nonlinear interaction of modes which we already encountered in Chapters IV and V; in this case the nonlinear interaction is considered very simply:

$$k^2_{\perp\max} = \frac{4k^2 u_{\max}}{l}. \tag{19.2}$$

Because of the exponential damping of the electromagnetic field outside the active part of the cross section, the diffraction light-energy losses connected with the radiation of the field into the surrounding space are a negligibly small quantity. This essential concept was first stated by Vainshtein [23].

The diffraction losses are practically nonexistent for all of the generated transverse modes which satisfy the inequality (19.1); therefore, no selection takes place among these modes.†

Such a simple consideration of the conditions at the boundary of the active portion of the cross section is possible only in the case of concave reflectors, for which the motion of the material point is localized in a potential well.

We now consider some particular cases.

§ 20. Cylindrical Mirrors

Let us consider mirrors in the shape of the surface of a cylinder (not necessarily a circular cylinder). In this case the equivalent mechanical system is a heavy material point in a smooth groove extending along the y axis. Assuming that the condition (18.18) is

† This clarifies the fact (strange at first glance) that the inequality (19.1), which determines the number of generating modes, does not contain the parameters of the active medium or the pumping which characterize the nonlinear interaction of the modes in a resonator having plane-parallel mirrors (Chaps. IV and V). The point is that the field distribution in a resonator with plane-parallel mirrors is determined by competition between the pumping and the diffraction phenomena. However, in the case of a resonator with concave reflectors the diffraction losses are so small that they cannot compete with the pump input, and therefore the pump intensity drops out of consideration.

satisfied, we first consider the motion of the material point from the classical standpoint.

In order for the material point to move within the confines of the active part of the cross section, its velocity must be directed parallel to the x axis. The material point performs oscillatory motion along this axis, and the maximum velocity of this motion is equal to $V = \sqrt{2\, g u_{max}}$, where $g = 2\, v^2/l$ is the acceleration of gravity in the mechanical system. Taking account of Eq. (18.7), we find the maximum angular divergence of the radiation in the xz plane:

$$\theta_{xz} = \frac{\sqrt{2gu_{max}}}{v} = 2\sqrt{\frac{u_{max}}{l}}. \tag{20.1}$$

Here u_{max} is the maximum value of u (i.e., the height of the mirror) within the confines of the active part of the cross section.†

Assume that the optic axes of the reflector form an angle (but their generants are parallel), i.e., each mirror is turned through a small angle $\vartheta/2$ about the y axis. Then the surface of the mirror is stipulated by the equation $z = u(x, y) + \vartheta x/2$. If $\vartheta < 2\, u'_{max}$, then z reaches a minimum in the active part of the cross section within which the material point performs oscillatory motion as previously (although with a smaller amplitude). This means that, as previously, light localized in the active part of the cross section can be generated, so that misalignment of the mirrors does not affect the generation threshold. From this we find the maximum allowable misalignment angle of the optic axes of the reflectors in the xz plane:

$$\vartheta_{max} = 2u'_{max}, \tag{20.2}$$

where u'_{max} is the maximum value of the derivative $u'(x)$ within the active part of the cross section.

For reflectors shaped in the form of the surface of a circular cylinder having the radius R, Eqs. (20.1), (20.2) take the form:

$$\theta_{xz} = a\sqrt{\frac{2}{Rl}}, \tag{20.3}$$

† The angular divergence of the radiation in the yz plane does not differ from the case of a resonator with plane mirrors (Chap. IV).

$$\vartheta_{max} = \frac{2a}{R}. \qquad (20.4)$$

These results, which have been obtained in the geometrical optics approximation, are applicable for the conditions (18.18), which can be written in the following form when Eq. (19.2) is taken into account:

$$u_{max} \gg \frac{l}{a^2k^2}. \qquad (20.5)$$

Taking the constraint (18.1) into account as well, we find the range of applicability of the given treatment:

$$\frac{D}{(4\pi)^2} \ll \frac{u_{max}}{\lambda} \ll \frac{1}{D}, \qquad D = \frac{\lambda l}{a^2} \ll 1. \qquad (20.6)$$

If the left inequality is substantially violated, then any noticeable difference from the case of a resonator with plane mirrors vanishes; a substantial violation of the right inequality leads to an abrupt growth of the diffraction losses, and the resonator does not sustain oscillation.

Naturally, the same results can also be obtained from the wave equations for conditions (18.18). Since we are interested in radiation in the xz plane, we shall seek the solution of Eq. (18.17) in the form $\psi(x)$; then the equation is written as follows:

$$\psi''(x) + \left[k_\perp^2 - \frac{4k^2u(x)}{l}\right]\psi = 0. \qquad (20.7)$$

In the range $|x| \ll a$, where the material point has its greatest velocity, it is necessary to place $u = u(0) = 0$ in the equation, and the solution has the form $\exp(ik_\perp x)$; from this the angular divergence is $\theta_{xz} = k_\perp/k$. Substituting (19.2) into this equation, we obtain Eq. (20.1) for the maximum angular divergence of the radiation.

§ 21. Reflectors in the Shape of a Surface of Rotation

Assume that the active part of the cross section has the shape of a circle, while the mirrors have the shape of an arbitrary concave surface of rotation which satisfies the condition (18.1). Let us introduce the cylindrical coordinates r, φ in the plane of the cross section with the origin at the center of the reflector.

For the sake of brevity we shall restrict our analysis to a classical treatment of the equivalent mechanical system. The velocity of the material point has a radial component V_r and a tangential component V_φ; correspondingly, it is necessary to consider two components θ_r and θ_φ of the angular divergence of the light. These quantities are determined by the largest possible values of V_r and V_φ for a material point whose motion is confined to the cross section of the sample.

A material point moving in the plane passing through the optic axis has the greatest radial velocity; under these conditions $V_\varphi = 0$, $V_{r\,max} = \sqrt{2\,gu_{max}}$. Taking (18.7) into account, we find the maximum radial angular divergence of the radiation:

$$\theta_r = \frac{V_{r\,max}}{v} = 2\sqrt{\frac{u_{max}}{l}}. \tag{21.1}$$

A material point moving in a circular orbit has the greatest tangential velocity, and here the centripetal force V_φ^2/r must be equal to the centripetal acceleration $gu'(r)$ (the mass of the material point is assumed to equal unity). From this we find the maximum angular divergence of the light:

$$\theta_\varphi = \frac{V_{\varphi\,max}}{v} = \sqrt{\frac{2}{l}\{u'(r)\,r\}_{max}} \tag{21.2}$$

(the subscript "max" designates the maximum value within the active part of the resonator cross section).

Figure 21 shows how the ray cross section increases during the same interval in the presence of radial (a) and tangential (b) angular divergences which are identical in magnitude. At large distances the diameter of the ray considerably exceeds the initial value; then the difference between θ_r and θ_φ vanishes, and it is sufficient to consider just the largest of these values.

By analogy with the case of cylindrical reflectors we find the maximum allowable misalignment of the mirrors:

$$\vartheta_{max} = 2u'(r)_{max}. \tag{21.3}$$

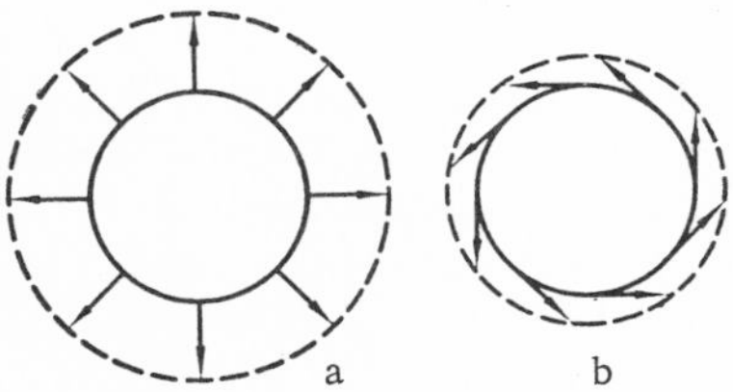

Fig. 21. The increase of the cross section of a generated ray during the same time interval in the presence of radial (a) and tangential (b) angular divergences of the same magnitude.

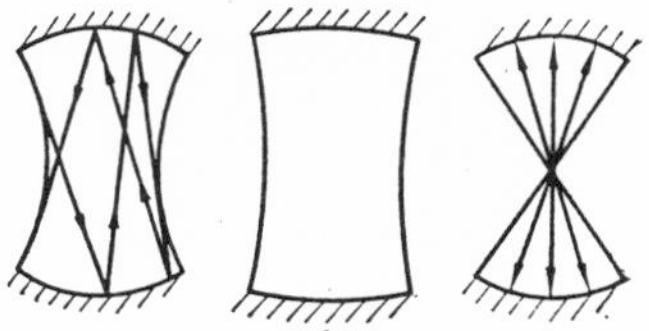

Fig. 22. Shape of the caustic surface which bounds the electromagnetic field in a spherical resonator for $R > l/2$ (a), $R \gg l/2$ (b), and $R = l/2$ (c).

What has been said above concerning a resonator with cylindrical reflectors remains valid so far as the applicability of this treatment is concerned.

Let us consider the important particular case of spherical reflectors in greater detail. Using R to designate the radius of curvature of the mirrors and a to designate the radius of the cross section of the active sample, we have the following result for the angular divergence:

$$\theta = \theta_r = \theta_\varphi = \frac{\sqrt{2}a}{\sqrt{Rl}}. \quad (21.4)$$

The maximum allowable misalignment angle of the reflectors is

$$\vartheta = \frac{2a}{R}. \quad (21.5)$$

The condition for the applicability of the treatment has the form

$$\left(\frac{D}{2\pi}\right)^2 \ll \frac{l}{R} \ll 1. \quad (21.6)$$

L. A. Vainshtein [23] considered resonators having spherical mirrors without the constraint expressed by the right inequality in (21.6); he showed that the electromagnetic field is concentrated inside the surface of a hyperboloid of rotation (a so-called caustic surface) and decreases exponentially beyond its boundaries (Fig. 22). This surface has a simple geometric meaning: it serves as the envelope of all the rays moving in the resonator. As the indices of the transverse modes increase, the caustic surface develops further and further away from the optic axis.

If the right inequality in (21.6) is satisfied, then the caustic surface is transformed into the surface of a circular cylinder whose cross section is the region of motion of a material point which is accessible to classical treatment. However, if R is decreased for a fixed l until $R > l/2$, then the angular and spatial distributions of the radiation are described qualitatively by the equations given above. For $R = l/2$ the field distribution undergoes a qualitative change (Fig. 22c). A further reduction of R, according to Vainshtein, leads to an abrupt growth of the diffraction losses.

Above we have neglected the refraction of light at the ends of the active element. With allowance for the refraction of light, the region in which the caustic exists takes the form [112]

$$R > R_c \equiv \frac{1}{2}\left[l - l_0\left(1 - \frac{1}{n}\right)\right] \tag{21.7}$$

where n is the refractive index of the active element. Let us give the equations for the angular divergence θ and the number of generated transverse modes $Z_\perp$, which are applicable throughout the entire range (21.7) for the condition $Z_\perp \gg 1$ [112]:

$$\theta = a_0[(l_0/2n)^2 + R_c(R - R_c)]^{-1/2} \tag{21.8}$$

$$Z_\perp = \frac{\pi^2}{3}\,\frac{a_0^4}{\lambda^2}\,\frac{R_c(R - R_c)}{[(l_0/2n)^2 + R_c(R - R_c)]^2} \tag{21.9}$$

(here a_0 is the radius of the generating region of the active element). The quantity (21.9) reaches a maximum for

$$R = R_c + l_0^2/4n^2R_c \tag{21.10}$$

Such a resonator configuration corresponds to the most stable operation with regular oscillations in accordance with §32 (provided only that the condition (21.10) is not violated due to thermal lens effects).

§ 22. The Spectral Composition of the Radiation

The frequency spectrum of the natural oscillations is stipulated by Eq. (18.15) in which $k_\perp^2$ takes the discrete values allowed by the Schrödinger equation (18.17). For example, in the case of spherical reflectors, for which $u(r) = r^2/2R$, the Schrödinger equation is satisfied by a Hermite function; we then have the following results for the natural frequencies:

$$\omega(m_3, m_\perp) = \frac{v}{l}\left[\pi m_3 + 2m_\perp \arcsin\sqrt{\frac{l}{2R}}\right]. \tag{22.1}$$

Here m_3 and $m_\perp$ are integer indices which characterize the motion of the radiation in the longitudinal and transverse directions, respectively.† The Vainshtein equation has been given here, since its applicability is not restricted by the right inequality (21.6).

† Only the index $m_\perp$ was included in Eq. (22.1), since degeneration occurs with respect to the indices of the transverse modes.

If the condition (18.18) is satisfied a large number of transverse modes are generated. These are satellites around each longitudinal mode, which is described by the second term in Eq. (18.15). Substituting (19.2), we find the spectral range of the transverse modes:

$$\delta\omega = \frac{vk_\perp^2}{2k} = \frac{2vku_{max}}{l}. \tag{22.2}$$

The ratio of this broadening to the interval between neighboring longitudinal modes is equal to

$$\frac{k_\perp^2 l}{2\pi k} = (k_\perp a)^2 \cdot \frac{D}{(2\pi)^2} = \frac{4u_{max}}{\lambda}. \tag{22.3}$$

This ratio is represented in the form of the product of two independent parameters – a large parameter and a small parameter; from this it follows that the ratio can vary within wide limits. If $u_{max} \ll \lambda/4$, then the longitudinal modes can be spectrally separated, and the radiation is generated in the form of equally spaced slightly broadened spectral lines; here the number of generated longitudinal modes is practically no different from the number in the case of a plane-parallel resonator. Therefore, the two systems of modes – longitudinal modes and transverse modes – do not interact with one another and can be treated independently as in the case of a plane-parallel resonator.

However, if $4u_{max} \gg \lambda$, we can no longer speak of the spectral lines of the longitudinal modes that have been broadened due to transverse modes. In fact, the frequency of a mode is determined to an equal degree by its longitudinal and transverse indices.† Therefore, for $4u_{max} \gg \lambda$ the interaction between the systems of transverse and longitudinal modes leads to a substantial rearrangement of the spectrum and the electromagnetic field structure.

Here we shall limit ourselves to a qualitative consideration of this phenomenon,‡ assuming for simplicity that in addition to the condition $4u_{max} \gg \lambda$ the relationship

$$\frac{4u_{max}}{\lambda} \gtrsim 2j \tag{22.4}$$

† In other words, the longitudinal modes are spectrally separated or unseparated, depending on the relationship between the wavelength and the angular deflection (i.e., the spread of the optical path length between reflectors); this result is quite natural.

‡ A quantitative investigation will be carried out in Chap. VIII.

is satisfied. Here 2j is the number of longitudinal modes generated in a plane-parallel resonator; this number is stipulated by Eq. (15.10). Condition (22.4) means that the spectral interval covered by the longitudinal modes (for a fixed transverse index) overlaps the interval covered by the transverse modes.

In Eq. (18.15) for the wave vector (which is proportional to the frequency) the first term depends on the longitudinal index of the mode, while the second term depends on its transverse indices. For the condition (22.4) both of these terms vary in approximately identical intervals. Because of this the longitudinal and transverse indices for each mode included in the generation process can be combined in such a way that the frequencies of all the modes fall within a narrow spectral band.† From what has been said above it is clear that under these conditions a large number ($\gtrsim$ 2j) of spectrally nonseparating longitudinal modes with different m_3 are generated; correspondingly, a practically uniform intensity distribution is established along the z axis. Once more we emphasize the fact that, unlike the case of plane reflectors (Chap. V), all of the generated longitudinal modes fit within a small spectral interval.

What has been said above can be clarified by means of Fig. 23 which illustrates the k-spaces. In this figure the parallel straight lines that are perpendicular to the z axis stipulate discrete values of k_3 (i.e., the wave vector must terminate on one of these straight lines). The wave vector forms an angle with the z axis which can vary within the limits of the angular spread of the radiation; the figure shows various directions of the wave vector. From the figure it is evident that a) the length of the wave vector has a small spread, and b) many overlapping longitudinal modes fit within the corresponding spectral interval.

An experimental investigation of the spectral composition of the radiation as a function of the geometric parameters of the resonator was carried out in [54] using a ruby laser. A more convenient resonator with plane-parallel mirrors and two identical positive lenses (having a focal length F) mounted in front of the mirrors inside the cavity was used instead of a spherical resonator. Such a

---

† The tendency to realize this minimum value of the spectral width is connected with the reduction of the gain with increasing distance from the central frequency ω_0. As will be shown in Chap. VIII, this minimum value is reached in the absence of other essential causes of spectral broadening.

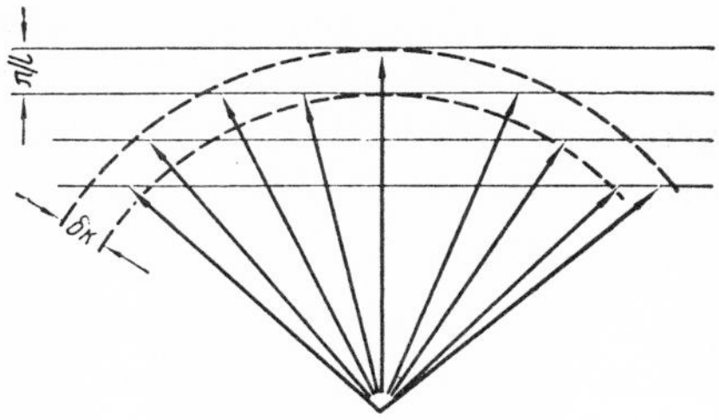

Fig. 23. The k-space (the parallel straight lines represent the $k_3 = \pi m_3 / l$ plane in the space of the wave vector $\vec{k}$; the arrows represent the values of $\vec{k}$ for the various modes).

resonator is equivalent to a spherical resonator having a radius of curvature R = F(§ 24). For a resonator with lenses the condition (22.4) takes the form

$$F < \frac{a^2}{\lambda j}. \qquad (22.5)$$

Experiment shows that in the case of weak lenses whose focal length exceeds the right side of (22.5) (under the conditions of the experiment described it is approximately 30 cm) the spectral width is not diminished relative to the case of a plane parallel resonator with no lenses. However, the use of sufficiently strong lenses which satisfy the relationship (22.5) leads to a substantial reduction of the spectral width (by approximately one order of magnitude).

§ 23. The Transition to a Resonator with Plane Mirrors

Let us trace the variation of the angular divergence of the laser radiation for a reduction of the radius of curvature of the reflectors, beginning with the value $R = \infty$ corresponding to plane mirrors. Since in the case of a resonator having plane mirrors the characteristics of the radiation are determined by the parameter $a\sqrt{\varkappa_1 k\zeta}$, it is convenient to consider the two limiting cases separately.

1. $a\sqrt{\varkappa_1 k\zeta} \gg 3$. For $R = \infty$ (plane mirrors) the number of generated transverse modes is a large quantity of the order of $a\sqrt{\varkappa_1 k\zeta}$. This leads to an angular divergence $\theta = 0.7\sqrt{\varkappa_1\zeta/k}$ of the radiation.

Let us introduce the characteristic value of the radius of curvature for which only the fundamental transverse mode of a spherical resonator is localized in the active part of the cross section having the radius a:

$$R^* = \frac{2k^2a^4}{l} = \frac{8\pi^2}{D^2}l. \qquad (23.1)$$

For a reduction of R from ∞ to R^* the angular divergence is not altered qualitatively, since the modes of the spherical resonator cannot be generated as yet. For $R < R^*$ the fundamental transverse mode is already generated, while for $R \ll R^*$ many transverse modes are generated, and the geometrical optics approximation becomes applicable.

For $R \ll R^*$ Eq. (21.4) for the angular divergence becomes applicable; this equation can be rewritten in the form

$$\theta = \frac{\lambda}{2\pi a}\sqrt{\frac{R^*}{R}}. \qquad (23.2)$$

From this it follows that if R^* does not greatly exceed R, then the angular divergence is comparable with the diffraction limit $\lambda/2a$. With allowance for what was said above, the dependence of the angular divergence on R has the form shown schematically in Fig. 24 (the solid line).

2. $a\sqrt{\varkappa_1 k\xi} \ll 1$. For $R = \infty$ (plane reflectors) the fundamental transverse mode of a resonator with plane-parallel mirrors is predominantly generated. When R is decreased to the value R^*, the fundamental transverse mode of a spherical resonator begins to be generated, and for a further decrease in R the higher transverse modes are generated also. The dependence of the angular divergence on R is shown by the dashed line in Fig. 24.

In [52] a reduction of the angular divergence of the light generated by a laser with long-focus mirrors was observed right up to the diffraction limit when the parameter R^* was reduced to a value comparable with the radius of curvature of the reflectors (the re-

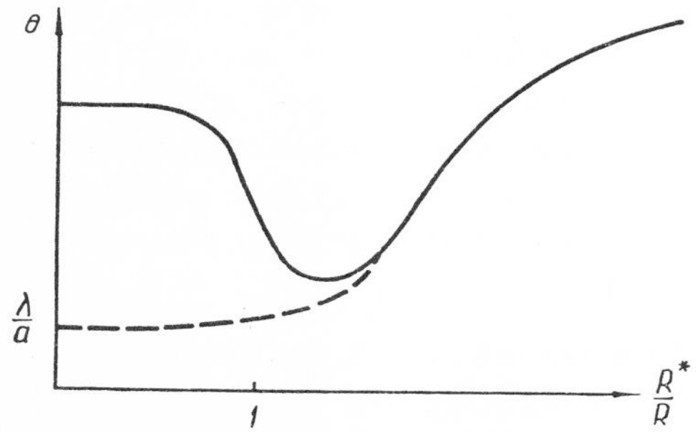

Fig. 24. Graphical representation of the dependence of the angular radiation divergence on the radius of curvature R of spherical reflectors.

duction in R^* was achieved by increasing the distance between reflectors).

§ 24. An Optical Cavity Resonator Containing Lenses

Let us consider an optical cavity containing one or two positive lenses which are placed in front of plane mirrors. If the path of the rays in such a resonator is considered, then it is not difficult to confirm the fact that a lens having the focal length F deflects rays in a manner which is similar to the deflection produced by a spherical mirror having a radius of curvature R = F (we consider the fact that for each reflection a ray passes through the lens twice). Therefore, a cavity with two identical lenses placed near plane reflectors is equivalent to a cavity with spherical mirrors having a radius of curvature R = F (however, if only one lens is placed inside the cavity, then it is necessary to place R = 2F).†

From the standpoint of the experimental procedure a cavity with lenses is very convenient, since experiments can be carried out with simple eyeglass lenses without aligning them carefully. Small errors of the lenses and their alignment (just as errors in the alignment of the mirrors) have practically no effect on the operation of the laser (see §20, 21).

In [73] an experimental investigation of the angular distribution of light in a cavity containing lenses was carried out. In accordance with what was said in § 21, it turned out that the angular distribution has fairly sharp boundaries. This makes it possible to introduce the total value of the angular divergence θ_{max}, whose experimental values are shown for lenses of various values of optical power by the points shown on the graph in Fig. 25. The straight line is drawn according to Eq. (21.4), where it is necessary to place R = F when two lenses are considered, and, the radius of the active part of the cross section must be assumed equal to 0.6 of the radius of the sample in accordance with § 6 (the sample had a smooth lateral surface). From Fig. 25 it is evident that the experimental data are in agreement with Eq. (21.4).

The use of lenses not only simplifies the experimental procedure, but also allows investigation of the optical errors of a cavity

† For condition (18.1) a cavity in which one mirror is plane and the other spherical is equivalent to a cavity with two spherical mirrors having half the curvature.

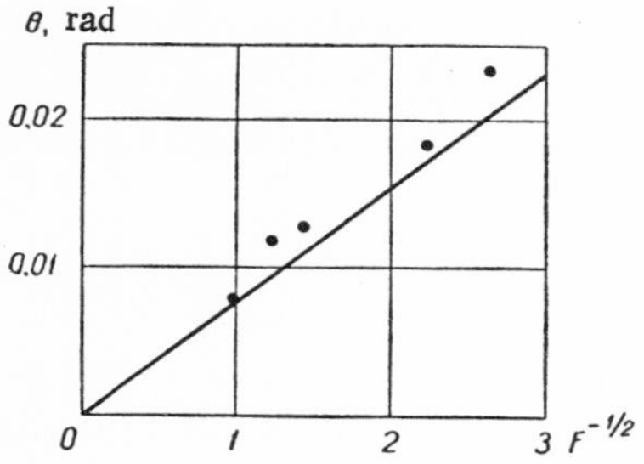

Fig. 25. Dependence of the angular divergence of the radiation in a cavity containing two lenses on the power of the lenses (expressed in diopters).

containing plane mirrors [74]. If a lens is placed in the cavity and the disalignment angle ϑ of the reflectors is increased continuously, then the generation threshold remains unchanged right up to a certain value ϑ_0, after which it begins to increase sharply. Figure 26 shows the dependence of ϑ_0 on the lens power 1/F. The experimental points fit the straight line $\vartheta_0 + \beta = \tilde{a}/F$, which does not pass through the origin. Here β is the deflection angle of a ray of the optical system for $\vartheta = 0$ (i.e., for parallel mirrors); in other words, β is the characteristic angle through which the sample deflects a ray passing through it. From Fig. 26 we find $\beta = 40''$, as well as $\tilde{a} = 1.5$ mm; evidently, $\tilde{a}$ is the radius of that region of the cross section in which generation develops for a very small excess of the pump power above threshold.

A lens can also be used to check the optical alignment of the system: if the introduction of the lens into the cavity lowers the generation threshold, then this means that the errors of the optical system affect the threshold power.

Further on (in Chap. VII) it will be shown that lenses placed in a cavity can also be used to obtain proper relaxation oscillations of the intensity.

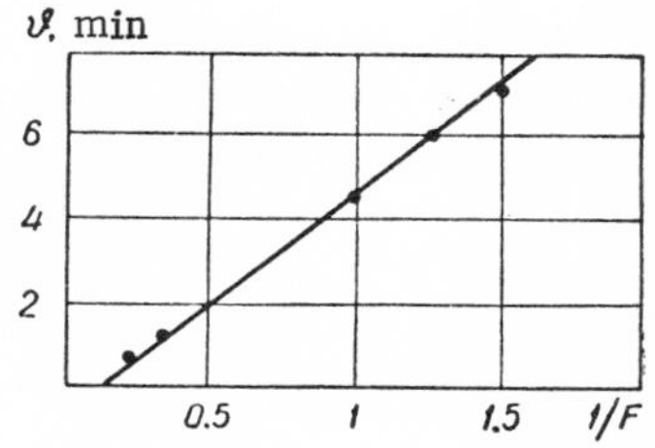

Fig. 26. Experimental dependence of the maximum allowable disalignment angle of the reflectors on the power of a lens placed in the cavity (F^{-1}, in diopters).

Finally, we note that due to thermal distortion of the active rod during the generation process a weak effective lens develops in the cavity; this lens is positive for air cooling and negative for water cooling [24].

Conclusions

1. The motion of the generated light in the plane of the cavity cross section is equivalent to the motion of a heavy material point along a concave upturned mirror surface. The geometrical optics approximation corresponds to motion of the material point according to the laws of classical mechanics, and wave optics corresponds to a quantum-mechanical treatment of the motion of the point.

2. If the radius of curvature of the mirror does not exceed the limits of the interval $l/2 < R \ll 2k^2 a^4/l$, then many transverse modes are generated simultaneously, and their diffraction losses are practically zero.

3. An angular divergence of the radiation equal to $2\sqrt{u_{max}/l}$, is connected with the generation of many transverse modes, where u_{max} is the height of the reflector within the active part of the cross section.

4. The longitudinal modes are spectrally separated if $u_{max} \ll \lambda/4$, and they are not separated in the converse case. If the longitudinal modes are separated, then the spectrum of the generated radiation is analogous to the case of plane reflectors (Chap. V), the only difference being that the spectral width of each longitudinal mode is a fraction $4u_{max}/\lambda$ of the interval between neighboring longitudinal modes, which is equal to $\pi v/l$. A sufficiently strong overlap of the longitudinal modes leads to a reduction of the spectral width of the generated radiation.

5. A cavity with plane mirrors and positive lenses is analogous to a cavity with spherical mirrors. The lenses which are placed in a cavity with plane mirrors coaxially with the active sample allow the degree of imperfection of the optical system to be estimated and its effect on the threshold pump power to be eliminated.

Chapter VII

RELAXATIONAL INTENSITY OSCILLATIONS

In this chapter we shall consider transients in a resonator containing a uniformly excited active medium. Homogeneity of the medium presupposes both uniformity of the pumping and uniformity of the intensity distribution of the generated light.†

As in the preceding chapters, our analysis will be carried out for a four-level diagram; however, all of the results can be generalized directly for the case of a three-level diagram using the relationships given in (9.3).

§ 25. The Kinetic Equation

Let us show that for the assumption of homogeneity of the active medium the kinetic equation is derived directly from the wave equation for the electromagnetic field. As above, we shall consider the effective absorption coefficient by introducing the complex dielectric constant $\varepsilon = \varepsilon_0 [1 + (i\varkappa/k)]$ where $\varkappa$ depends on time but not on the coordinates. We write the wave equation for the magnetic field [36]:

$$\Delta\vec{H} - \frac{1}{v^2}\left(1 + \frac{i\varkappa}{k}\right)\frac{\partial^2\vec{H}}{\partial t^2} = 0, \quad v = \frac{c}{\sqrt{\varepsilon_0}}. \tag{25.1}$$

We shall asume $\varkappa$ to be a stipulated time function which varies substantially during a time interval t_0 that considerably exceeds all the remaining characteristic times of the problem (in particular, ω^{-1}). This allows us to seek the solution in the form of an almost monochromatic wave, which justifies writing the wave equation (25.1) without considering the spectral dependence of the complex dielectric constant.

† Oscillations in the output of lasers were studied in [132, 133, 134, 135, 136].

If we place $\vec{H} = \vec{h}(\vec{r})f(t)$, then in Eq. (25.1) the variables are separated and $k^2 \equiv \omega^2/v^2$. For the spatial part we obtain the conventional wave equation $\Delta\vec{h} + k^2\vec{h} = 0$, while for the time part we have

$$\left(1 + \frac{i\varkappa}{k}\right)\frac{d^2f}{dt^2} + \omega^2 f = 0, \tag{25.2}$$

We shall seek the solution of Eq. (25.2) in the form of an almost monochromatic wave:

$$f = A(t)e^{-i\omega t}, \tag{25.3}$$

where the amplitude A changes noticeably only over a time t_0. Substituting into the equation and neglecting the terms which are quadratic in the small parameters t_0^{-1} and $\varkappa/k$, we have

$$\dot{A} + \frac{v\varkappa}{2}A = 0. \tag{25.4}$$

From this we obtain

$$A(t) = A(0)\exp\left[-\frac{v}{2}\int_0^t \varkappa(t)\,dt\right]. \tag{25.5}$$

Thus, the time dependence of the intensity of the almost monochromatic field components under investigation has the form

$$J(t) = J(0)\exp\left[-v\int_0^t \varkappa(t)\,dt\right]. \tag{25.6}$$

Let us now consider nonmonochromatic light. In this case it is necessary to write the analogous wave equation with the dielectric-constant operators [36] instead of Eq. (25.1). However, since such an equation is formally linear, the superposition of almost monochromatic components of the type investigated here satisfies it. These components have different frequencies, and therefore different values of the quantity† $\varkappa = \varkappa(\omega, t)$ correspond to them.

The intensity of the resultant field, which is proportional to the time-averaged square of the field modulus, is equal to the sum

† The quantity $\varkappa(\omega, t)$ has meaning for sufficiently large t_0, for which the inequality $\tilde{\omega} \gg 1/t_0$ is satisfied; here $\tilde{\omega}$ is the frequency interval, and t_0 is the time interval during which $\varkappa$ varies substantially.

of the intensities of the monochromatic components, since the products of fields having different frequencies drop out during the process of time averaging.† Thus, the spectral energy density of nonmonochromatic light depends on time as follows:

$$\varrho(\omega, t) = \varrho(\omega, 0) \exp\left[-v \int_0^t \varkappa(\omega, t)\, dt\right], \tag{25.7}$$

or, in differential form,

$$\frac{\partial \varrho(\omega, t)}{\partial t} = -v\varkappa(\omega, t)\,\varrho(\omega, t). \tag{25.8}$$

Equation (25.8) expresses the fact that the light, in propagating in the active medium at the velocity v, traverses the path $v$dt during the time dt and is attenuated by the amount $\rho\varkappa v$dt. We could have written equation (25.8) originally but then we would not have a criterion of its applicability.

Equation (25.8) does not consider that portion of the spontaneous emission which is retained in the resonator and is in the direction of the stimulated emission. Let us write the power of this portion of the spontaneous emission in the form $N^*\delta\Omega/4\pi$; here N^* is the threshold pump power, which with an accuracy of up to the luminescence energy yield (which we have assumed equal to unity) is equal to the spontaneous-emission power (see § 6), while $\delta\Omega$ is a parameter whose order of magnitude coincides with the solid angle within which the stimulated radiation is distributed. When spontaneous emission is included, the equation for the spectral density of the generated light takes the form

$$\frac{\partial \varrho(\omega, t)}{\partial t} = -v\varkappa(\omega, t)\,\varrho(\omega, t) + \frac{\delta\Omega}{4\pi} N^* P(\omega) \tag{25.9}$$

(the factor $P(\omega)$ describes the spectral distribution of the spontaneous emission).

For a four-level diagram the effective absorption coefficient (4.3) can be written in the form

$$\varkappa(\omega) = \varkappa_1 - Bn\frac{P(\omega)}{P(\omega_0)} = \varkappa_1 - \varkappa_1 \frac{n}{n^*} \cdot \frac{P(\omega)}{P(\omega_0)}. \tag{25.10}$$

† Here we do not consider the interference between modes having different frequencies. In [12] it was shown that interference between modes whose frequency difference is less than the quantity T^{-1} (see §30) is substantial. For a ruby or neodymium–glass laser the value of T is large enough so that we could neglect this interference.

Here B is a constant which is determined from the condition that the generation threshold is reached for $\varkappa(\omega_0) = 0$; in this case the number of excited atoms n is equal to its threshold value n^*. Let us expand the function $P(\omega)$ describing the shape of the luminescence band into a series in powers of the frequency $\omega' = \omega - \omega_0$:

$$P(\omega) = P(\omega_0)\left[1 - \frac{\omega'^2}{\overline{\omega}^2}\right]. \tag{25.11}$$

Here $\overline{\omega}$ is a quantity of the order of the half-width of the luminescence band; for a Gaussian band $\overline{\omega} = 0.6\ \Delta\omega$.

Taking account of (25.11), we write Eq. (25.10) as follows:

$$\varkappa(\omega) = \varkappa(\omega_0) - \varkappa_1 \frac{\omega'^2}{\overline{\omega}^2}, \tag{25.12}$$

$$\varkappa(\omega_0) = \varkappa_1\left(1 - \frac{n}{n^*}\right). \tag{25.13}$$

In Eq. (25.12) we limited the expression to quadratic terms of the expansion in ω', since in view of the smallness of the spectral width of the laser radiation compared with $\overline{\omega}$ the investigation was restricted to the range $|\omega'| \ll \overline{\omega}$. This allowed us similarly to neglect the small difference between n and the threshold value n^* in the quadratic term of the expansion (in § 28 we shall show that the relative magnitude of this difference is stipulated by the small parameter $\sqrt{\xi}/\sqrt{v\varkappa_1 T}$).

Substituting the expansion (25.12) into Eq. (25.9), we reduce this equation to its final form:

$$\frac{\partial\varrho(\omega, t)}{\partial t} = -v\varrho(\omega, t)\left[\varkappa(\omega_0, t) - \varkappa_1\frac{\omega'^2}{\overline{\omega}^2}\right] + \frac{\delta\Omega}{4\pi}N^*P(\omega). \tag{25.14}$$

So far we have assumed $\varkappa$ to be a stipulated time function. In order to obtain a closed system of equations in $\rho(\omega, t)$ and $\varkappa$ it is necessary to supplement Eq. (25.14) with the equation which describes the dependence of the quantity $\varkappa(\omega_0)$ (i.e., actually n) on the light energy:

$$\frac{dn}{dt} = \frac{N}{\hbar\omega_0} - \frac{n}{T} - \frac{v\varkappa_1 n}{n^*\hbar\omega_0}J; \quad J = \int \varrho(\omega, t)\, d\omega. \tag{25.15}$$

The first term in the right side is the number of atoms excited by the pumping per unit time; the second term is the number of atoms which emit spontaneously per unit time, and the third term describes the stimulated emission of the excited atoms.

As we have already noted, the threshold power is equal to the spontaneous-emission power for $n = n^*$; i.e., $N^* = n^* \hbar\omega_0 / T$. Using this fact and replacing the quantity n by the value n^* which is close to n in the right side of Eq. (25.15), we write the equation in the more convenient form

$$\frac{T}{n^*}\cdot\frac{dn}{dt} = \zeta - \frac{v\varkappa_1}{N^*}J;\quad J = \int \varrho(\omega, t)\, d\omega. \tag{25.16}$$

Equations (25.14) and (25.16) form a closed system with allowance for Eq. (25.13). In certain cases it is convenient to combine these equations into one equation; for this purpose Eq. (25.14) is solved for $\rho(\omega, t)$:

$$\varrho(\omega, t) = \frac{\delta\Omega}{4\pi} N^* P(\omega_0) \int_0^t \exp\left[-y(t) + y(t') - v\varkappa_1(t - t')\frac{\omega'^2}{\overline{\omega}^2}\right] dt'. \tag{25.17}$$

Here we have introduced the substitution

$$y(t) = v\int_0^t \varkappa(\omega_0, t')\, dt',\quad \dot{y} = v\varkappa(\omega_0, t). \tag{25.18}$$

Using Eq. (25.13), n is easily expressed in terms of $\dot{y}$:

$$n = n^*\left(1 - \frac{\dot{y}}{v\varkappa_1}\right) \tag{25.19}$$

Let us integrate the quantity (25.17) with respect to frequency and substitute the result into Eq. (25.16). Expressing $\dot{n}$ in terms of $\dot{y}$ in this equation, we reduce it to the form

$$\frac{\ddot{y}}{v\varkappa_1} + \frac{\zeta}{T} - \frac{v\varkappa_1}{T}\cdot\frac{\delta\Omega}{4\sqrt{\pi}} P(\omega_0)\,\overline{\omega} \int_0^t \frac{e^{-y(t)+y(t')}}{\sqrt{v\varkappa_1(t - t')}} dt'. \tag{25.20}$$

Converting to dimensionless time:

$$u = t\sqrt{\frac{v\varkappa_1\zeta}{T}}, \tag{25.21}$$

we obtain the kinetic equation in final form [66]:

$$-\frac{d^2y}{du^2} = 1 - \alpha e^{-y(u)} \int_0^u e^{y(u')} \frac{du'}{\sqrt{u - u'}}, \tag{25.22}$$

where

$$\alpha = \frac{\delta\Omega}{4\sqrt{\pi}} \cdot \frac{N^*}{N - N^*} \left(\frac{v\varkappa_1 T}{\zeta}\right)^{1/4} P(\omega_0)\,\bar{\omega} = \frac{\delta\Omega}{4\pi} \cdot \frac{N^*}{N - N^*} \left(\frac{v\varkappa_1 T}{\zeta}\right)^{1/4} \quad (25.23)$$

(the latter equation applies to a Gaussian luminescence band).

The quantity α, which is proportional to the small solid angle $\delta\Omega$, is the small parameter of the theory and (speaking very roughly) stipulates the ratio of the energy of the stimulated emission in the interval between intensity peaks to its maximum value.

The condition for applicability of the kinetic equation (25.22) is a uniform distribution of the pumping intensity and the stimulated emission intensity in the active medium (see § 29 for greater detail).†

Equation (25.22) describes oscillations of the stimulated-emission intensity. Before we investigate this problem mathematically, let us examine it qualitatively.

§ 26. The General Picture of the Oscillations of the Stimulated Emission

The existence of emission oscillations derives from simple physical concepts. Assume that the pumping is switched on at time $t = 0$, after which its intensity remains fixed. At first there is no stimulated emission in the resonator, and the number of excited atoms increases under the influence of the pumping (Fig. 27a). At a certain time $t = t_1$ the number of excited atoms reaches the threshold value n^*, after which the effective absorption coefficient $\varkappa(\omega_0)$ becomes negative, and generation of stimulated emission begins in the resonator. The initial stimulated-emission is triggered from a very small fraction of the spontaneous-emission energy which falls within the small solid angle $\delta\Omega$. Therefore, at values of t slightly exceeding t_1 the stimulated part of the energy in the resonator is still very small, its emitting action is insignificant, and the number of excited atoms continues to increase because of the pumping. However, the stimulated-emission energy increases exponentially with time, and at a certain time $t = t_2$ its emitting action begins to predominate over the pumping action; beginning at

† More detailed criteria for the applicability of the kinetic equation are given in the monograph by V. S. Mashkevich [51].

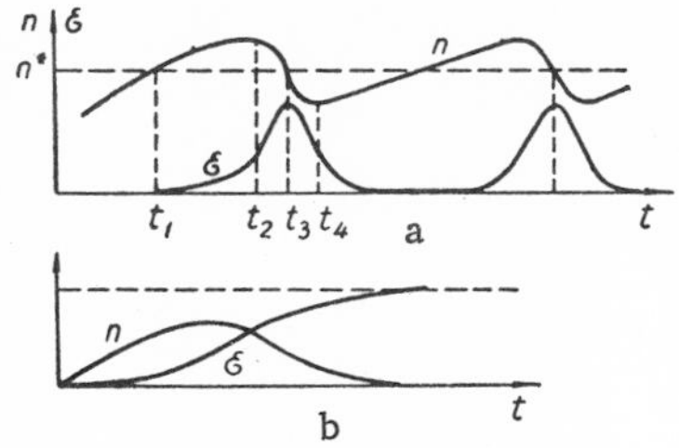

Fig. 27. Plot of the dependence of the stimulated-emission energy $\mathscr{E}$ and the number of excited atoms n as functions of time for a four-level laser: a) $\varkappa_1 \neq 0$; b) $\varkappa_1 = 0$.

this instant, the number of excited atoms will decrease with time (Fig. 27a). However, n does not decrease to the threshold value immediately, but only by the time t_3. But in the interval between t_2 and t_3 the number of excited atoms exceeds the threshold value, so that $\varkappa < 0$, and the light energy continues to increase. For $t > t_3$, n drops below the threshold value, $\varkappa$ becomes positive, and the light energy begins to decrease; its maximum is obviously reached at time t_3 when $n = n^*$. For $t > t_3$, the number of excited atoms also decreases simultaneously with the light energy, since this energy is still close to its maximum value and its emitting action predominates, as previously, over the pumping. This continues until time t_4, when the light energy decreases so much that the probability of stimulated emission of excited atoms becomes lower than the probability of excitation of the atoms by the pumping. For $t > t_4$, n again begins to increase, and the light energy, as previously, decreases and becomes very small. Then the next period of oscillations begins. The magnitude of the period is determined by the relatively low rate of replenishment of the number of excited atoms which have emitted, and the duration of the intensity peak is determined by its high rate of rise, which is proportional to the light velocity; from this it is clear that the intensity peak covers only a small fraction of a period, i.e., intensity oscillations having a large amplitude must develop in the resonator.

Note that the existence of the described oscillations requires that both n and the light energy $t_3 < \dot{\mathscr{E}} < t_4$ decrease simultaneously in a certain time interval $t_3 < t < t_4$. This is possible only as a result of light losses which are proportional to $\varkappa_1$. Actually, for $\varkappa_1 = 0$ the generation threshold for a four-level diagram would be zero, and the conditions $\varkappa(\omega_0) < 0$, $\dot{\mathscr{E}} > 0$ would always apply. In this case the light energy would be a nondecreasing time function, and the relaxation would not be accompanied by oscillations, as shown in Fig. 27b.

It is clear that for a sufficiently small $\varkappa_1$ the relaxation of the system will similarly be accompanied by a monotonic variation of $\mathscr{E}$. As $\varkappa_1$ (and therefore n^* also) increases, the value of n can decrease further and further below its threshold value, and the amplitude of the relaxation oscillations must increase accordingly. Below we shall show that the damping of the oscillations actually does increase with a decrease in $\varkappa_1$.†

§ 27. Oscillations Having a Large Amplitude

The kinetic equation (25.22) can describe continuous operation when $\varkappa(\omega_0)$ = const, as well as a slight deviation from continuous operation, which is accompanied by small intensity oscillations. The equation for small oscillations has a simple form. It was investigated in [66, 92, 103], but we shall not dwell on this in detail since in the experiment the generation is accompanied by very strong oscillations rather than small ones.

The kinetic equation is also simplified in the limiting case of strong intensity oscillations. Let us recall that Eq. (25.22) is the balance equation for the number of excited atoms n. The left side of the equation is proportional to $\dot{n}$, and in the right side the first term describes spontaneous emission from atoms and their excitation by the pumping (the pump intensity is assumed fixed), while the second describes stimulated emission from the excited atoms.

Below it will be shown that the solution of the kinetic equation with a large oscillation amplitude has the form of a slowly varying periodic function (Fig. 28); here the emission is generated in the form of sharp peaks (the periods and the peaks associated with them will be numbered with the subscript k). The integral term in Eq. (25.22), which is proportional to the light energy in the resonator, must be considered only in the region of the peaks (indicated on Fig. 28 by the solid bars). Outside this region the integral terms can be dropped, and the equation can be integrated in elementary fashion. We have

$$y(u) = -\frac{1}{2}(u - u_k)^2 + y(u_k). \tag{27.1}$$

† This may seem somewhat unexpected, since usually the damping of the oscillations is connected with the energy dissipation, which in this case is proportional to $\varkappa_1$. In reality, however, the damping of the oscillations of light energy is not connected with the light-energy losses, which are compensated for by the pumping, but by another, finer mechanism.

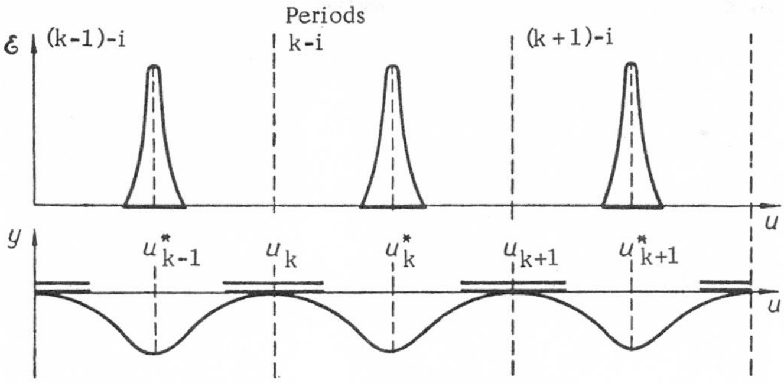

Fig. 28. Diagram for deriving the equation for oscillations having a large amplitude: the upper curve shows the intensity of the stimulated emission, and the lower curve shows the logarithm of the intensity with a reversed sign.

Here u_k is the left boundary of the k-th period (i.e., the point located midway between the (k − 1)-th and k-th peaks (Fig. 28)). In choosing the integration constants we considered the fact that the solution of (27.1) must be symmetrical relative to the point u_k (Fig. 28).

The integral term in Eq. (25.22) plays a role only in the region of the peaks; at the same time only the vicinity of the points u_k makes a substantial contribution to the integral (this vicinity is isolated by a double bar in Fig. 28). Actually, it follows from Eq. (27.1) that the integrand decreases exponentially with departure from the point u_k.

If in Eq. (25.22) the independent variable is outside the region of the peaks, then the integral term can simply be dropped. However, if u belongs to the region around the k-th peak, then the integral is actually taken over the vicinity of the points $u_k, u_{k-1}, \ldots,$ which precede the investigated peak, so that the integral term has the form

$$\alpha e^{-y} \sum_{p=1}^{k} \int_{-\infty}^{u} \exp\left[y(u_p) - \frac{(u' - u_p)^2}{2}\right] \frac{du'}{\sqrt{u - u'}} \qquad (27.2)$$

(in view of the exponential decrease of the integrand, the integration can be extended to infinite limits).

Note that in fact u varies within the limits of a narrow peak with its center at the point u_k^* in Eq. (27.2), while u' varies in the vicinity of the points u_p. This allows us to place $u = u_k^*$, $u' = u_p$ in the multiplier of the exponential. Now the integral is easily calculated, and the integral term in Eq. (25.22) takes the form†

$$\alpha \sqrt{\pi} \sum_{p=1}^{k} e^{y(u_p)-y} (u_k^* - u_p)^{-1/2}. \tag{27.3}$$

Now Eq. (25.22) can be written in the following fashion during the k-th period:

$$\frac{d^2 y_k}{du^2} + 1 - Y_k \exp(-y_k - Y_k) = 0. \tag{27.4}$$

Here we have introduced a notation for the k-th period whose advantages will be revealed further on:

$$y_k = y - y(u_k), \tag{27.5}$$

$$Y_k e^{-Y_k} = \alpha \sqrt{2\pi} \sum_{p=1}^{k} e^{y(u_p)-y(u_k)} (u_k^* - u_p)^{-1/2}. \tag{27.6}$$

The quantity Y_k varies slowly from period to period. In this connection the solution of the equation for the oscillations can be split into two problems: first it is necessary to consider the equations during one period for fixed Y, and then it is required to find the law for the damping of the oscillations (i.e., for the decrease in Y which accompanies an increase in the period number).

In this section we shall consider the first part of the problem. In conformity with this we fix the number of the period and rewrite Eq. (27.4) omitting the subscript k:

$$\frac{d^2 y}{du^2} = -\frac{dU}{dy}, \tag{27.7}$$

where

$$U = y + Ye^{-Y-y}. \tag{27.8}$$

According to the notation in (27.5), $y \equiv y_k$ is measured from its maximum value during the k-th period, so that

$$y_{\max} = 0 \tag{27.9}$$

---

†The integral term has this form for the condition that u does not exceed the limits of the k-th peak; however, in fact Eq. (27.3) is applicable throughout the entire k-th period. In reality, both the integral term in Eq. (25.22) and Eq. (27.3) which approximates it are exponentially small outside the peak region.

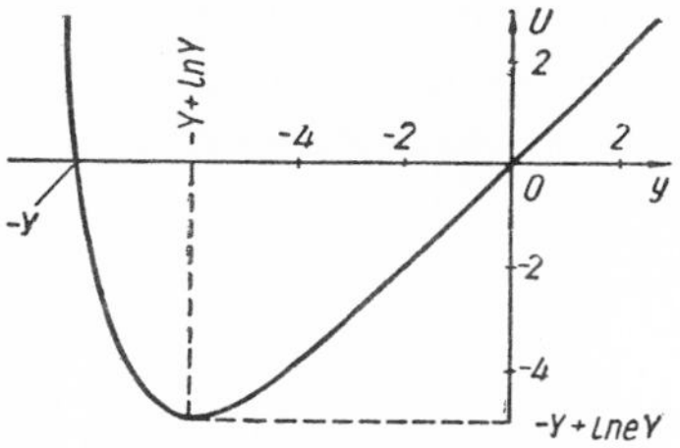

Fig. 29. Graph of $U = U(y)$ for $Y = 8$.

In deriving Eq. (27.7) we assume that the oscillations have a sufficiently large amplitude and that the intensity peaks are sufficiently sharp. Only for such a condition can we isolate the vicinity of the point u_k^*, where the emission intensity is high, from the vicinity of the point u_p, where the integrand in Eq. (25.22) is large. From what follows it will be evident that Y is the amplitude of the oscillations, and the condition for applicability of the analysis has the form

$$Y \gg 1. \tag{27.10}$$

Below it will be shown that for conventional resonators the condition (27.10) is satisfied; this is verified by the experimental data.

We have reduced the kinetic equation to the form of the equation for mechanical oscillations of a material point having a unit mass, which has been placed in a potential field U (Fig. 29). The amplitude of these oscillations is fixed by condition (27.9), from which it follows that the motion of the material point is bounded by the region $-Y < y < 0$ isolated by the thick solid line in Fig. 29. Therefore, the quantity $Y = y_{max} - y_{min}$ is none other than the amplitude of the oscillations.

The potential field in which the material points move approaches asymptotically close to the straight line $U = y$ for large y, and increases exponentially for $y \to -\infty$. When the material point performs oscillations it moves relatively slowly along the right flat branch of the potential curve and considerably faster along the left steep branch (Fig. 29). The left branch corresponds to an intensity peak during which the reserve of excited atoms is rapidly depleted due to the action of powerful emission, and the right branch corresponds to the interval between peaks, when the reserve of excited atoms is slowly replenished due to the pumping. As the amplitude Y increases, the difference between the slopes of the left and right branches of the potential curve increases, and their intensity peak becomes sharper.

The equation for the mechanical oscillations (27.7) can be integrated conventionally. First, considering y as the independent variable, we obtain the first integral of the equation which expresses the energy conservation law of the material point:

$$\frac{1}{2}[y'(u)]^2 + U(y) = 0. \tag{27.11}$$

We place the total energy of the material point equal to zero in accordance with the condition (27.9) while neglecting the exponentially small quantity Y exp (-Y).

Separating the variables in Eq. (27.11) and carrying out still another integration, we find the solution of the equation for the oscillations (27.7) in implicit form:

$$u - u_k^* = \pm \frac{1}{\sqrt{2}} \int\limits_{-Y}^{y} \frac{dy'}{\sqrt{-y' - Ye^{-Y-u'}}} \tag{27.12}$$

(the "minus" sign is chosen for the first half-period, while the "plus" sign is chosen for the second).

From Eq. (27.12) we find the period of the oscillations:

$$\Delta u \equiv u_{k+1} - u_k = \sqrt{2} \int\limits_{-Y}^{0} \frac{dy}{\sqrt{-y - Y\exp(-Y-y)}} \tag{27.13}$$

The dependence of the period on the amplitude is shown graphically in Fig. 30.

The function y(u) stipulated by Eq. (27.12) can be tabulated for various values of the amplitude Y. Figure 31 shows the time dependence of the stimulated-emission energy, which has the follow-

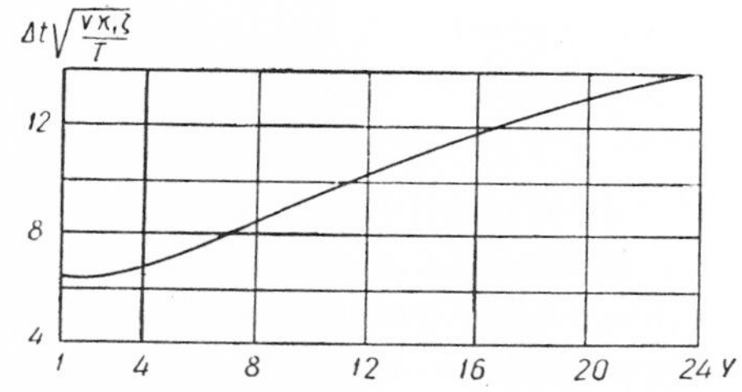

Fig. 30. Dependence of the period Δt of free relaxation oscillations of the intensity on the amplitude Y of the oscillations.

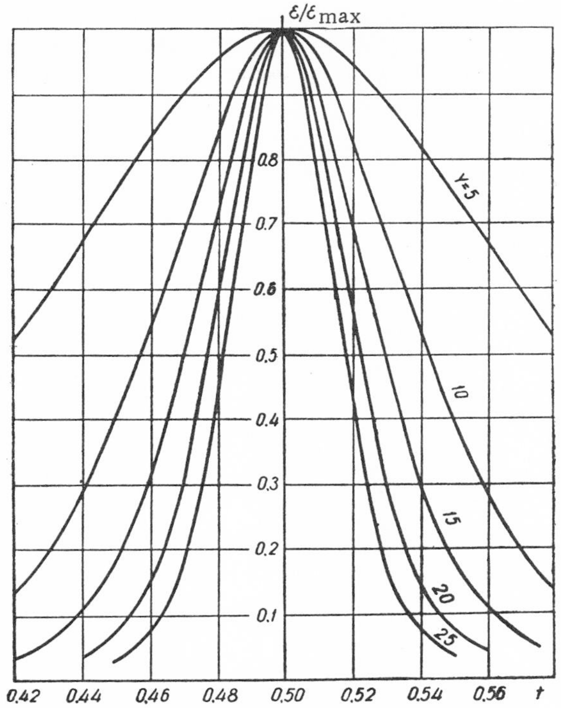

Fig. 31. The shape of the peak of the intensity oscillations for various values of the amplitude Y of the oscillations (the time in fractions of a period is plotted along the axis of abscissas).

ing form within a certain constant factor:

$$\mathscr{E} = Ye^{-Y-y} \quad (-Y \leqslant y \leqslant 0). \tag{27.14}$$

From the figure it is evident that the width of the peak is small compared with the period (t is expressed in period units, and the maximum ordinate of the peaks is assumed to be unity).

It is not difficult to obtain an approximate analytical solution of Eq. (27.7) in the peak region, where the first term under the radical in Eq. (27.12) describing the pumping plays a small role in comparison with the exponential second term, which describes the stimulated emission. Therefore, in the peak region it is possible to replace the first term by its value Y corresponding to the intensity maximum. Then the integral in (27.14) is calculated in elementary fashion, and finally the expression for the intensity takes the form

$$\mathscr{E} = Ye^{-Y-y} = \frac{Y}{ch^2 \sqrt{\frac{Y}{2}}(u - u_k^*)} . \qquad (27.15)$$

From this it is evident a) that Y is the height of the peak, expressed in certain units, and b) that outside the peak region the intensity decreases exponentially.

From Eq. (27.15) we derive the asymptotic expression for the half-width δt:

$$\delta t = \sqrt{\frac{T}{v\varkappa_1\zeta}}\,\delta u = \sqrt{\frac{T}{v\varkappa_1\zeta}}\left(\frac{2{,}49}{\sqrt{Y}} + O\left(\frac{1}{Y^{3/2}}\right)\right). \qquad (27.16)$$

We also give the asymptotic expression for the oscillation period:

$$\Delta t = \sqrt{\frac{T}{v\varkappa_1\zeta}}\,\Delta u = \sqrt{\frac{T}{v\varkappa_1\zeta}}\left(\sqrt{8Y} + O\left(\frac{1}{\sqrt{Y}}\right)\right). \qquad (27.17)$$

The ratio between the half-width of the peak and the period of the oscillations is

$$\frac{\delta t}{\Delta t} = \frac{0.88}{Y} + O\,(Y^{-2}) \qquad (27.18)$$

and vanishes asymptotically as the amplitude Y increases.

§ 28. The Damping of the Oscillations and Their Amplitude

In the left side of Eq. (27.4) we drop the small terms which lead to damping of the oscillations. These terms have the form

$$\sqrt{\frac{\zeta}{v\varkappa_1 T}}\frac{dy_k}{du}\left(\frac{1}{\zeta} + Y_k \exp\left(-Y_k - y_k\right)\right) -$$

$$- \alpha \int_0^{u_k^*} [(u - u')^{-1/2} - (u_k^* - u')^{-1/2}]\, e^{y(u') - y(u)}\, du'. \qquad (28.1)$$

The first term is small because of the factor $\sqrt{\zeta / v\varkappa_1 T} \lesssim 0.01$ (usually $T \sim 10^{-3}$ sec), while the second term is small due to the factor $a \sim \delta\Omega$.

Let us clarify the origin and meaning of Eq. (28.1). In deriving the equations for the oscillations we neglected the small difference between n and the threshold value n^*; this is equivalent to neglecting the quantity $\varkappa(\omega_0)$ compared with $\varkappa_1$. Under these conditions a relative error of the order of $\varkappa(\omega_0)/\varkappa_1 = \dot{y}/v\varkappa_1 = y_u'\sqrt{\xi/v\varkappa_1 T}$ is allowed, which is corrected by the first term in Eq. (28.1). The deviation of n from the threshold value n^* has a negative sign during the first half-period in which the emission in the resonator increases; because of this the height of the peak decreases somewhat compared with that of the previous peak. Conversely, during the second half-period in which the intensity decreases, $n > n^*$; this facilitates the decrease of the dip between peaks. The damping mechanism connected with this, which is described by the first term in Eq. (28.1), is called the n-damping mechanism for the sake of brevity.

Then, in calculating the integral in Eq. (27.2) we place $u = u_k^*$ in the multiplier in front of the exponential. The second term in Eq. (28.1) represents the difference between the exact and approximate values of this integral and depends explicitly on time; this leads, as we know, to a change in the amplitude of the oscillations (in this case to a reduction). This damping is connected with the fact that the exponential term in the equations for the oscillations (the second term in (28.1) is a correction to this term) takes account of the spontaneous emission retained in the resonator within the solid angle $\delta\Omega \sim Y\exp(-Y)$, as well as the stimulated emission. But the spontaneous emission which is added to the stimulated emission has a different spectral composition. It is this which leads to the given damping mechanism, which we shall call the ω-mechanism for the sake of brevity.

In order to calculate the damping it is simplest to interpret Eq. (27.7) with the small terms of (28.1) as the equation for the oscillations of a material point in the presence of small dissipative forces. It is not difficult to calculate both the change in the energy of the material point during a period under the influence of these forces and the related change in the magnitude of y. In this way we obtain the equations which relate the difference $y(u_{k+1}) - y(u_k)$ to Y_k. In all we therefore have k equations with 2k unknowns $y(u_1), \ldots, u(u_k), Y_1, \ldots, Y_k$. It is possible to use the definition (27.6) as the missing system of k equations. In principle, the calculation is simple but rather cumbersome; therefore we shall give only the final result. For greater clarity we shall assume that either the first or the second of the mechanisms investigated is dominant.†

For the case in which the n-damping mechanism predominates the amplitude Y, which is proportional to the height of the peak, depends on the number k of the peak as follows:

$$Y_k = \left[\frac{1}{\sqrt{Y_1}} + \frac{\sqrt{8}}{3}\sqrt{\frac{\zeta}{v\varkappa_1 T}}\left(1 + \frac{1}{\zeta}\right)k\right]^{-2}. \tag{28.2}$$

However, if the ω-mechanism predominates, then the height of the peak varies with the peak number according to the law

$$Y_k = Y_1 - \ln\left\{\frac{Y_1}{Y_k}\left(1 + \frac{3k}{\sqrt{\pi}}\right)\right\}. \tag{28.3}$$

Here Y_1 is the initial value of Y and has the following form in accordance with (27.6):

$$Y_1 = \ln\frac{1}{\alpha}\sqrt{\frac{\Delta u}{4\pi}} + \ln Y_1 \cong \ln\frac{1}{\alpha} + \frac{3}{2}\ln\ln\frac{1}{\alpha} - 0.2 \tag{28.4}$$

(we used Eq. (27.17) for the period). Because of the smallness of the α, the quantity Y_1 is fairly large (under usual experimental conditions $Y \gtrsim 10$), so that the condition (27.10) is satisfied.

The damping rate can be characterized conveniently by the number of peaks $k_{1/2}$ during which their height decreases by a factor of two. For the n- and ω-damping mechanisms we have the following results, respectively:

$$k_{1/2}^{(n)} = 0.44\sqrt{\frac{v\varkappa_1 T}{Y_1}}\frac{1}{\zeta^{1/2} + \zeta^{-1/2}}, \tag{28.5}$$

$$k_{1/2}^{(\omega)} = \frac{\sqrt{\pi}}{3}\left[\frac{1}{2}\exp\left(\frac{Y_1}{2}\right) - 1\right]. \tag{28.6}$$

If the first of these expressions is less than the second, it can be assumed that the n-damping mechanism described by Eqs. (28.2) and (28.5) is dominant; in the converse case Eqs. (28.3) and (28.6) must be used.

---

† In a number of papers, for example [92, 103], the spectral distribution of the intensity (and therefore the ω-mechanism) is neglected.

From Eqs. (28.5), (28.6) it is evident that with an increase in the solid angle $\delta\Omega$, which determines the initial amplitude Y_1, the n-damping mechanism becomes less effective, while the ω-mechanism is intensified.

Usually, the n-mechanism does not have time to produce continuous-wave operation during a pumping pulse. Continuous-wave generation can be obtained by means of more powerful damping mechanisms.

For a sufficiently large divergence angle (of the order of several degrees) the ω-mechanism leads to damping during several periods.

There also exists a damping mechanism which is connected with the curvature of the caustic surface [28]. For almost concentric mirrors or almost confocal lenses† this mechanism leads to practically continuous-wave generation.

Finally, powerful damping and practically continuous-wave operation can be achieved through the use of a passive modulator having a negative modulating capacity (see § 48 for greater detail).

§ 29. The Range of Applicability and the Experimental Data

The conclusion drawn above concerning the existence of proper relaxation oscillations is based on the assumption that the active medium is spatially uniform.

In the case of a plane-parallel resonator this assumption includes the assumption of a uniform distribution of the pump intensity and uniform intensity of the refractive index of the active medium (the relative magnitude of the nonuniformities must not exceed $\lambda/2l \sim 10^{-5}$). Moreover, the uniformity of the active medium requires the generation of a large number of modes. This condition is not always satisfied for longitudinal modes (see end of § 15), while the simultaneous generation of a large number of transverse modes requires satisfaction of the inequality (11.1). In any case, the production of proper relaxation oscillations in a plane-parallel resonator requires the satisfaction of rigorous conditions: uniform

† The position of the foci must be determined while taking account of the refraction of the light at the end faces of the active element.

pump distribution and uniformity of the refractive index (the latter condition is usually violated during the degeneration process as a result of nonuniform heating of the active medium). If no special measures are adopted in the experiment to assure careful satisfaction of these conditions, then a disordered pattern of intensity peaks representing the superposition of unmatched intensity oscillations of individual modes or of portions of the active medium which are subjected to different conditions is observed instead of proper damped oscillations.†

It is considerably simpler to obtain proper oscillations of the intensity in a resonator having spherical mirrors with a sufficiently large curvature, or in a plane-parallel resonator with sufficiently powerful positive lenses. A relatively large angular divergence leads to averaging of the nonuniform pump distribution over the resonator cross section; therefore, the condition requiring uniform pumping drops out.

Further, assume that the height u_{max} of a spherical reflector within the limits of the active region is fairly large (in any case, it must exceed λ considerably). Then a small nonuniformity of the refractive index plays no role, provided only that its magnitude is small compared with the ratio u_{max}/l_0 (under these conditions the spread of the optical path length between mirrors, which is connected with the nonuniformity of the refractive index, is small compared with the height u_{max}).

Finally, the condition $u_{max} \gg \lambda$ provides for the generation of a large number of transverse modes, equal to $k_\perp a$; this is evident from Eq. (22.3), which can be rewritten in the form

$$(k_\perp a)^2 = \left(\frac{16u_{max}}{\lambda}\right)\left(\frac{\pi^2}{D}\right),$$

where $D \ll 1$. As far as the longitudinal modes which go over into generation are concerned, their number is similarly fairly large (for the condition (22.4) or (22.5) it exceeds the number of generated

† The case in which the irregularity of the oscillations is connected with the generation of a small number of separable modes was investigated experimentally in [30, 31]. These papers establish a correspondence between the modes and the oscillogram peaks; on the basis of experimental data it was concluded that (in conformity with the presentation above) the condition for proper oscillations is the simultaneous generation of many modes under approximately identical conditions.

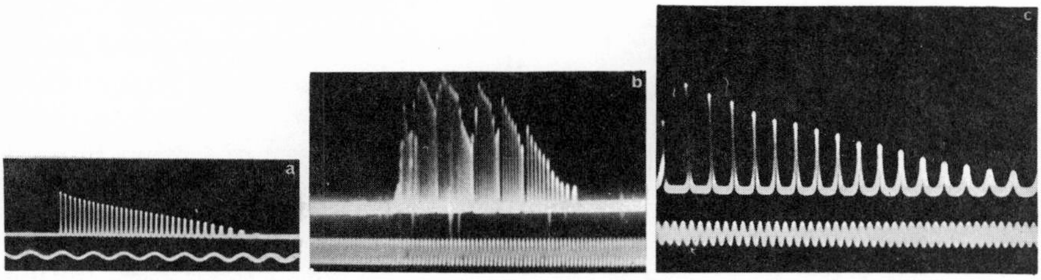

Fig. 32. Oscillograms of the intensity oscillations: a) two identical lenses (F = 14.4 cm) and good ruby; b) the same, but poor ruby; c) poor ruby with two pairs of lenses (F_1 = 20 cm, F_2 = 60 cm). Sweep frequency: a) 20 kHz; b) 90 kHz; c) 200 kHz.

longitudinal modes of a plane-parallel resonator; see §31, subsection 2).

Thus, proper relaxation oscillations are generated in a spherical resonator, provided only that the height $u_{max} = a^2/2R$ of the reflector is sufficiently large.† This similarly applies to a plane-parallel resonator containing lenses; here the role played by the curvature is played by the quantity $u_{max} = a^2/2F$.

What has been said can be interpreted from a somewhat different, clearer point of view. The kinetic equation describing regular oscillations of the radiation is an energy relationship which does not take account of the wave properties of light. The latter are nonessential in the case of a sufficiently large number of generated modes, just as a quantum-mechanical system becomes classical if it occupies a sufficiently large number of states.

Regular damped oscillations in a spherical resonator were observed by a number of authors (for example, [75, 91]).

A plane-parallel resonator with lenses is more convenient than a spherical resonator in the sense of experimental procedure, since the lenses do not require a precise alignment and can easily be shifted within the resonator (or their number and power can be changed). These parameters can be chosen in such a way that the intensity oscillations become completely regular [54]. As an example, Fig. 32 shows oscillograms from [54]. Figure 32a shows the oscillogram of the oscillations obtained in a resonator with two

† For the quantitative criterion see Chap. VIII.

identical lenses (F = 14.4 cm) which were placed near the reflectors, and a ruby rod having a relatively high optical performance; Fig. 32b shows the oscillogram under the same conditions except for a rod having poorer optical performance. From Fig. 32b it is evident that optical inhomogeneity of the active rod leads to instability of the proper oscillations, and as a result these oscillations decay into trains. In order to obtain a completely regular pattern of oscillations with the same rod it is necessary to alter the configuration of the resonator elements in such a way that a larger number of generated transverse modes (21.9) results. The corresponding oscillogram is shown in Fig. 32c.

§ 30. Self-Oscillations

As was shown above, in the case of a spatially uniform active medium the intensity oscillations are damped, so that continuous operation is stable.† However, in the case of a nonuniform active medium continuous operation can turn out to be unstable, and this means that generation must be accompanied by self-oscillations. In particular, the random undamped oscillations which accompany generation in a conventional plane-parallel resonator that is optically too imperfect are evidently self-oscillations of such origin.

As Mashkevich [50, 51] showed, instability of continuous operation can be caused by nonuniformity of the active-medium connected with the presence of various types of optical centers.

Self-oscillations are also possible in other cases in which the analysis used in this chapter becomes inapplicable: in particular, in the case of modes having close frequencies whose interference cannot be neglected as it was in deriving the kinetic equation in § 25. In [11, 12] the interaction of two modes having close frequencies (differing by less than T^{-1}) was considered; this interaction is caused by the fact that the phases of the modes remain matched for the duration of the lifetimes of an atom in the excited states. For a sufficiently small spectral interval between modes this interaction leads to instability of continuous-wave operation (i.e., to self-oscillations).

---

† This has been verified by the conclusions drawn by Livshits and Tsikunov [45] concerning the damping of small oscillations of a large number of longitudinal modes (§16) whose generation leads to practical uniformity of the active medium.

In the paper by Suchkov [76] it was shown via numerical integration of the transient nonlinear wave equation which takes the nonuniformity of the pumping into account that self-oscillatory operation exists in a certain range of laser parameters.

Let us consider the simple example of self-oscillations which are connected with misalignment of the plane end reflectors of a laser. If the optic axes of the mirrors form a small angle ϑ, then an estimate based on the geometrical optics approximation† shows that any ray exits from the resonator after a time

$$t_0 = \frac{4}{v}\sqrt{\frac{al}{\vartheta}} \tag{30.1}$$

has elapsed. This estimate is applicable for sufficiently large ϑ.

Assume that the misalignment angle ϑ is large enough so that the estimate (30.1) is at least qualitatively applicable. Then it can be assumed that if at a certain instant in time u' (we make use of the dimensionless time scale introduced in § 25) emission develops in the resonator, then at the subsequent instant u the fraction of this emission retained in the resonator is equal to $g(u - u')$, where $g(u)$ is a certain function which is close to unity for $u < u_0$ and decreases rapidly for $u > u_0$. Taking this into account, we write Eq. (25.22) in a somewhat modified form:

$$-\frac{d^2y}{du^2} = 1 - \alpha e^{-y}\int_0^u e^{y(u')}\frac{g(u-u')}{\sqrt{u-u'}}\,du'. \tag{30.2}$$

Equation (30.2), just as (25.22), can describe continuous-wave operation. In order to investigate the stability of continuous-wave operation it is necessary to consider small oscillations of the intensity and to determine the sign of the imaginary part of their frequency. As a result of simple calculations it turns out that for the case in which the time interval u_0 (during this time interval the function g

† For one reflection the angle φ formed by the direction of the ray with the optic axis changes by ϑ (i.e., $d\varphi/d\nu = \vartheta$, where ν is the number of reflections). From this we have $\varphi = \vartheta\nu$, and the coordinate of the reflection point of the ray is $l\int\varphi d\nu = l\vartheta\nu^2/2$. Obviously, the ray exists from the resonator after it has been displaced through the distance $2a$ in the transverse direction (a is the radius of the cross section); here the distance $2a$ is traversed by the ray in the two directions. Finding the number of reflections from this and multiplying it by l/ν, we obtain Eq. (30.1).

differs substantially from zero) is close to a small integer number of oscillation periods the amplitudes of the small oscillations increase exponentially with time (i.e., continuous-wave operation is unstable).

Without dwelling in detail on the case of small oscillations, let us consider the more interesting case of self-oscillations having a large amplitude [70]. For simplicity we shall assume that u_0 coincides with the period.

Above it was shown that the integral in Eq. (30.2) actually is taken over the vicinities of the points u_0 which cause the function y to have a maximum (Fig. 28). Let us consider Eq. (30.2) in the region of one of these intensity peaks; the instant corresponding to maximum intensity is designated by u^*. The nearest of the points u_p is separated from the instant u^* by a half-period; the next point is separated by 1.5 periods, etc. (Fig. 28). In calculating the integrals appearing in Eq. (30.2) it is sufficient to consider only the closest of these points; the remaining points u_p do not contribute to the integral because of the cutoff multiplier $g(u^* - u_p)$. An integral similar to the one considered here was calculated in § 27 and was given in the form of the sum (27.6). In the present case it is sufficient to retain just one term in this sum — the term which corresponds to the closest of the points u_p, which is a half-period distant from the instant u^*. Thus, the integral turns out to be equal to $2\sqrt{\pi}/\Delta u$ (Δu is the period of the oscillations), and Eq. (30.2) takes the form

$$\frac{d^2y}{du^2} + 1 - 2\alpha\sqrt{\frac{\pi}{\Delta u}}\,e^{-y} \equiv \frac{d^2y}{du^2} + 1 - Ye^{-Y-y} = 0. \qquad (30.3)$$

This equation coincides with Eq. (27.4), which was investigated in detail above; the only difference lies in the fact that the amplitude is fixed:

$$Y = \text{const} \cong \ln\frac{1}{\alpha} + \frac{3}{2}\ln\ln\frac{1}{\alpha} - 0.2. \qquad (30.4)$$

Physically, the fact that the amplitude is fixed can be related to the fact that the light energy cannot be retained for long in a resonator with misaligned mirrors; therefore each intensity peak develops due to the spontaneous-emission energy amplified by the active medium during the last half-period preceding the given peak, rather than due to the stimulated-emission energy which has accumulated previously in the resonator; during such a short time the emission cannot exit from the resonator. Thus, the self-oscillations are a train of disconnected intensity peaks, and since the peaks are formed independently, and under identical conditions, their amplitudes are identical.

Here we shall consider a resonator with plane mirrors which has a small angular spread of the radiation (usually this spread does not exceed several minutes). From this it follows that the amplitude of the oscillations is large ($Y > 15$) and that the radiation is generated in the form of very narrow peaks whose width is approximately one twentieth the width of a period.

The presentation above is in good agreement with experimental data. Experiment shows that when the mirrors are misaligned by an angle of one–two minutes the nature of the oscillations is altered qualitatively: the peaks become considerably sharper, the period increases, and the oscillations become regular [70].

Conclusions

1. For the case in which the radiation in the resonator behaves as a single whole its intensity undergoes proper oscillations which are characterized by an amplitude Y proportional to the height of the intensity peaks. The period, shape, and height of the peaks, as well as the damping rate of the oscillations, are expressed directly in terms of the amplitude. For $Y \gg 1$ the radiation is generated in the form of narrow peaks covering a small portion of the period.

2. The initial value of the amplitude is $Y_1 \sim |\ln \delta\Omega| \gg 1$; Y decreases with time, and the oscillations die out. Several damping mechanisms exist; certain of them can be reduced to rapid damping of oscillations and practically continuous-wave generation.

3. Experimental observation of proper damped oscillations in a plane-parallel resonator requires the satisfaction of rigorous conditions which provide for spatial uniformity of the active medium. However, in the case of spherical reflectors this requires only a sufficiently large radius of curvature of the reflectors within the limits of the generating region cross section (see also § 31).

4. In certain cases self-oscillations are possible which are related to instability of continuous-wave operation.

Chapter VIII

THE SPECTRAL WIDTH OF THE RADIATION

In this chapter we investigate the spectral composition of laser radiation while taking account of the most essential physical causes of spectral broadening. One of these causes — the nonlinear interaction between modes — was considered above (Chap. IV – VI). Below we also investigate the effect of angular divergence and kinetic operation on the spectral width of the generation.

The treatment in this chapter is carried out for a four-level diagram and small ζ, but the results can be generalized for the case of a three-level diagram using the substitutions in (9.3), and for the case of finite ζ by making use of the transformation $\zeta \to \zeta/(\zeta + 1)$.

§ 31. Spectral Composition of the Radiation for Continuous-Wave Operation

Let us investigate the spectral composition of the generated light in the limiting case of small and large angular divergence.

1. The Case of a Small Angular Divergence. Let us begin by examining the case in which the number of generated transverse modes is so small that the spectral interval covered by them is much smaller than the interval between neighboring longitudinal modes (Fig. 33b). Under these conditions the longitudinal structure of the electromagnetic field considered in § 15 is basically retained. The spectrum of the generated radiation consists of equidistant lines having the frequencies $\pi m v/l$, each of which can have several adjacent transverse modes with nonzero indices (Fig. 33b). The spectral interval $\delta\omega$ covered by all of the generated longitudinal modes is stipulated by Eq. (15.11) and thus does not depend on the angular divergence. In other words, the longitudinal modes practically do not interact with the transverse modes.

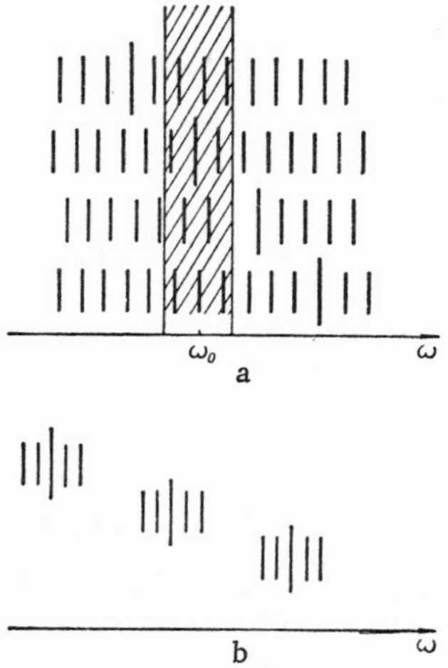

Fig. 33. Diagram of the structure of the generation spectrum (each row corresponds to a fixed longitudinal index): a) the case of a large divergence (the band of generated frequencies is shaded); b) the case of a small angular divergence.

If not only the fundamental but also the higher transverse modes are generated, then the spectral interval covered by them (for a fixed longitudinal index) can be related to the magnitude of the angular divergence θ of the light. Making use of the smallness of θ, we have $k_\perp \leq k\theta$ for the transverse component of the wave vector; however, the longitudinal component is $k_\parallel = \pi m/l$ (m is the longitudinal index). From this we find the magnitude of the wave vector

$$k = \frac{\omega}{v} = \sqrt{k_\parallel^2 + k_\perp^2} \cong k_\parallel + \frac{k_\perp^2}{2k} \leqslant \frac{\pi m}{l} + \frac{k\theta^2}{2}. \tag{31.1}$$

The last term in this equation corresponds to the spectral interval $\delta\omega_\perp$ covered by the transverse modes, whence

$$\delta\omega_\perp = \frac{\omega_0\theta^2}{2}. \tag{31.2}$$

In particular, for a plane-parallel resonator we have the following result for the condition (11.1) (i.e., for strong interaction of the transverse modes) when Eq. (11.5) is considered:

$$\delta\omega_\perp = \frac{v\varkappa_1\zeta}{4}. \tag{31.3}$$

In the case of a resonator containing concave mirrors having a small curvature we have $\theta = 2\sqrt{u_{max}/l}$ in accordance with Chap. VI. Substituting this result into Eq. (31.1), we find

$$\delta\omega_\perp = \frac{2\omega_0 u_{max}}{l}, \tag{31.4}$$

which is in agreement with Eq. (22.3).

Equations (31.3)-(31.4) have meaning for the condition

$$\frac{\lambda}{4a} \ll \theta < \sqrt{\frac{\pi v}{l\omega_0}}. \tag{31.5}$$

This means that at least several transverse modes must be generated, but at the same time the spectral range covered by them must be less than $\pi v/2l$.

2. Large Angular Divergence. In this case the spectral interval covered by the generated transverse modes considerably exceeds the interval between neighboring longitudinal modes (Fig. 33a). A substantial spectral overlap of the groups connected with different longitudinal modes occurs. From Fig. 33a it is evident that, unlike the previous case, the generated modes cannot be split into two weakly interacting systems of longitudinal and transverse modes. Let us show that the interaction between the longitudinal and transverse modes leads to a substantial reduction of the spectral width of the generation. Physically, this is explained by the fact that the frequencies of a large number of modes fall within a small spectral interval in the neighborhood of the maximum of the amplification band.

It can be shown that the spectral width of the generated light is determined basically by the angular divergence. This makes it possible to perform calculations for a plane-parallel resonator with a large angular divergence (caused, for example, by complete internal reflection from smooth walls of the active element) and to generalize the results for the case of a resonator having an arbitrary shape [67].

In the case of a closed plane-parallel resonator with a rectangular cross section the intensity of an individual mode has the form

$$\mathcal{E}_{mpq} = A_{mpq} \sin^2 \frac{\pi m z}{l} \sin^2 \frac{\pi p x}{a} \sin^2 \frac{\pi q y}{a}. \tag{31.6}$$

Here a is the side of the cross section; A_{mpq} is the positive amplitude of the intensity (m is the longitudinal index; p and q are transverse indices). Making use of Eq. (31.1) or (7.5), we find the spectrum of the modes:

$$\frac{\omega_{mpq} - \omega_0}{\bar{\omega}} = \frac{\sqrt{\xi}}{8} [\alpha (m - m_0) + \beta (p^2 + q^2)], \tag{31.7}$$

where

$$\alpha = \frac{8\pi v}{l\bar{\omega}\sqrt{\zeta}}, \quad \beta = \frac{4\pi^2 v^2}{a^2\omega_0\bar{\omega}\sqrt{\zeta}}. \tag{31.8}$$

(The longitudinal index m_0 corresponds to the center frequency ω_0 for $p = q = 0$; writing the spectrum in the form (31.7) turns out to be convenient later on.) Under the usual experimental conditions $\alpha \ll 1$, $\beta \ll 1$.

The equations for the amplitudes A_{mpq} can be obtained from the condition for continuous operation using the same method as that used in the case of an infinite cross section (§ 15). This equation is conveniently represented in the following form:

$$A_{mpq} = \gamma_{mpq} - [\alpha(m - m_0) + \beta(p^2 + q^2)]^2, \tag{31.9}$$

where

$$\gamma_{mpq} = 64 - 8\sum_{m'p'q'} A_{m'p'q'} - 4\Big(\sum_{p'q'} A_{mp'q'} + \sum_{m'q'} A_{m'pq'} + \sum_{m'p'} A_{m'p'q}\Big) - 2\Big(\sum_{q'} A_{mpq'} + \sum_{p'} A_{mp'q} + \sum_{m'} A_{m'pq}\Big). \tag{31.10}$$

For simplicity we shall assume that the angular divergence of the light is bounded by a cone which is coaxial with the laser; then the region of summation over the transverse indices has the form

$$p > 0, \quad q > 0, \quad p^2 + q^2 \leqslant \bar{p}^2. \tag{31.11}$$

Since we are considering multimode operation, it follows that each summation index in Eq. (31.10) takes a large number of different values. Since under these conditions positive intensity amplitudes appear under the summation sign, it is possible to neglect the double and single sums, as well as the quantities $\gamma_{mpq} \sim A_{mpq}$, in comparison with the triple sum. Taking this into account and substituting (31.9) into (31.10), we obtain the following equation:

$$8 - \sum_{mpq} A_{mpq} = 8 - \sum_{mpq}\{\gamma_{mpq} - [\alpha(m - m_0) + \beta(p^2 + q^2)]^2\} = 0. \tag{31.12}$$

From Eq. (31.7) it is evident that if the longitudinal indices do not exceed the limits of the interval

$$m_0 - \frac{\beta\bar{p}^2}{\alpha} \leqslant m \leqslant m_0, \tag{31.13}$$

it follows that the difference $\omega_{mpq} - \omega_0$ can be made to vanish by the appropriate choice of the transverse indices from the allowed range (31.11). Therefore, the range (31.13) contains a large number of modes whose frequencies are close to the maximum ω_0 of the amplification band. It can be shown that these modes form a narrow peak of the spectral intensity distribution. The remaining modes, which do not belong to the range (31.13), appear considerably further away from the point ω_0 on the frequency scale and form broad symmetrical wings of the spectral distribution.

We shall assume that the integrated intensity of the peak considerably exceeds the integrated intensity of the wings. Then in Eq. (31.12) it is possible to limit the summation to the range (31.13), while the solution can be sought in the form $\gamma_{mpq} = \text{const} = \gamma$. The summation in Eq. (31.12), which extends over a large number of positive terms, can be replaced by integration with respect to the continuous variables m, p, q. Under these conditions the range of integration with respect to m is stipulated by the inequality (31.13), while the range of the integration with respect to p and q is determined from the conditions requiring a nonnegative integrand and has the form

$$p \geqslant 0, \quad q \geqslant 0, \quad p^2 + q^2 \leqslant \frac{\alpha}{\beta}(m_0 - m). \tag{31.14}$$

The integral can easily be calculated in cylindrical coordinates and turns out to equal $\pi\overline{\varphi}^2\gamma^{3/2}/3a$. From this we find the solution of Eq. (31.12):

$$\gamma = \left(\frac{24\alpha}{\pi\overline{p}^2}\right)^{2/3}. \tag{31.15}$$

Substituting the solution (31.15) into Eq. (31.9), we find the intensity distribution with respect to the modes in the region of the peaks

$$A_{mpq} = \left(\frac{24\alpha}{\pi\overline{p}^2}\right)^{2/3} - \frac{64}{\zeta}\left(\frac{\omega_{mpq} - \omega_0}{\overline{\omega}}\right)^2 \tag{31.16}$$

Taking account of the fact that A_{mpq} is positive, we calculate the width of the peak using (31.16):

$$\delta\omega_p = \frac{\sqrt{\zeta\omega}}{2}\left(\frac{3\alpha}{\pi\overline{p}^2}\right)^{1/3} = \frac{\delta\omega_{\parallel}}{(\pi\overline{p}^2)^{1/3}} \sim \delta\omega_{\parallel}\left(\frac{\lambda}{\theta a}\right)^{2/3}. \tag{31.17}$$

where the substitutions in (15.11) have been used. With an accuracy within a numerical factor, Eq. (31.17) is also applicable to a spherical resonator. Taking (21.4) into account, we have

$$\delta\omega_p \sim \delta\omega_{\parallel} \sqrt[3]{\frac{lR}{k^2a^4}}. \tag{31.18}$$

Using analogous calculations, it is possible to find the spectral width of the waves and the ratio between the integrated intensities of the wings and the peak:

$$\delta\omega_W = \delta\omega_{\parallel} \sqrt{\frac{2\delta\omega_{\parallel}}{3\delta\omega_{\perp}}} \sim \delta\omega_{\parallel} \sqrt{\frac{\delta\omega_{\parallel}}{\theta^2\omega_0}}, \tag{31.19}$$

$$\frac{\mathscr{E}_W}{\mathscr{E}_p} = \left(\frac{2\delta\omega_{\parallel}}{3\delta\omega_{\perp}}\right)^{3/2} \sim \frac{1}{\theta^3}\left(\frac{\delta\omega_{\parallel}}{\omega_0}\right)^{3/2}. \tag{31.20}$$

Here we have introduced the spectral interval

$$\delta\omega_{\perp} = \frac{1}{2\omega_0}\left(\frac{\pi v \bar{p}}{a}\right)^2, \tag{31.21}$$

covered by all generated transverse modes.

For a spherical resonator we have

$$\frac{\mathscr{E}_W}{\mathscr{E}_p} \sim \left(\frac{\delta\omega_{\parallel}}{\omega_0}\cdot\frac{lR}{a^2}\right)^{3/2}, \qquad \frac{\delta\omega_W}{\delta\omega_{\parallel}} \sim \sqrt{\frac{\delta\omega_{\parallel}}{\omega_0}\cdot\frac{lR}{a^2}}. \tag{31.22}$$

From Eqs. (31.19) and (31.20) it is evident that substantial narrowing of the generation spectrum relative to the narrowing in the case of a plane-parallel resonator occurs for a sufficiently large divergence angle:

$$\theta > \sqrt{\frac{\delta\omega_{\parallel}}{\omega_0}}, \tag{31.23}$$

for which the spectral interval $\delta\omega_{\perp}$ covered by the generated transverse modes exceeds the spectral width of the radiation from a plane-parallel resonator (thereby the qualitative concepts stated in § 22) are confirmed). For a spherical resonator the condition for substantial narrowing of the spectrum has the form

$$\frac{a^2}{lR} > \frac{\delta\omega_{\parallel}}{\omega_0}. \tag{31.24}$$

The inequality (31.23) or (31.24) is simultaneously the condition for the applicability of the results obtained.†

The narrowing of the generation spectrum which has been considered is physically related to the fact that the peak of the spectral distribution is formed by degenerate modes whose frequencies practically coincide with the maximum ω_0 of the amplification band. As the number of such degenerate modes increases, the peak becomes more pronounced, and the ratio (31.20) becomes smaller. On the other hand, the degree of degeneration is characterized by the size of the interval $\delta\omega_\perp$. For $\delta\omega_\perp > \delta\omega_\parallel$, almost all of the longitudinal modes overlap spectrally (i.e., they degenerate) so that almost the entire generated intensity is incorporated in the peak.

The condition (31.23) or (31.24) for a substantial narrowing of the spectrum is also the condition for proper kinetic operation. Actually, since the overwhelmingly larger portion of the generated intensity is incorporated in the numerous degenerate modes with close amplitudes and different spatial intensity distributions, it follows that the distribution of the total intensity (and therefore the distribution of the population inversion) turns out to be practically uniform. What has been said can be interpreted more clearly. The degenerate modes, which have closely spaced frequencies, have amplitudes whose magnitudes are close and vary with time according to a practically identical law; it is this which provides for regular kinetic operation.

The above can be substantiated by experimental data. In [54] the narrowing of the spectrum and proper kinetic operation were observed simultaneously when sufficiently powerful lenses satisfy-

† Relationships (31.17)-(31.24) are inapplicable near the boundary of the region in which the caustic exists (i.e., for R close to the value R_c determined by Eq. (31.7)). In this case we have

$$\delta\omega_p \sim \frac{\delta\omega_\parallel}{Z_\perp^{1/3}}, \qquad \delta\omega_w \sim 0.4\,\delta\omega_\parallel \frac{(2j)^{1/2}}{Z_\perp^{1/4}}, \qquad \frac{\xi_w}{\xi_p} \sim 0.1\,\frac{(2j)^{3/2}}{Z_\perp^{3/4}}$$

Here $Z_\perp$ and 2j are the numbers of the transverse and longitudinal modes stipulated by Eqs. (21.9) and (15.10), respectively. The criterion (31.23) or (31.24) is replaced by the inequality

$$Z_\perp > \frac{1}{3}\,j^2 \equiv 0.17\left[\frac{\zeta}{\zeta+1}\left(\frac{l\bar{\omega}}{\pi v}\right)^2\right]^{2/3}$$

ing the conditions (31.24) were introduced into the cavity. For a fixed focal distance and fixed positions of the lenses the satisfaction of the criterion (31.24) depends on the diameter of the generating portion of the active element. If the generated portion of the cross section is reduced to a sufficient extent by means of an iris, then narrowing of the spectrum ceases to be observed, and the proper kinetic operation is simultaneously violated.

A similar experiment in kinetic operation of a laser (without explanation or comparison with the spectra) was given in [93], where a gradual transition from regular operation to random operation was observed when the generating portion of the rod in a spherical resonator was reduced.

3. Consideration of the Spectral Broadening Connected with Spontaneous Emission. The physical cause of this broadening resides in the fact that the laser radiation contains spontaneous-emission (luminescence) energy which is retained in the resonator and amplified by the active medium; here this spontaneous emission has a relatively large width.† The contribution of the spontaneous emission to the spectral width of the generation is determined by the equation [60, 97]

$$\delta\omega_{sp} \sim \frac{\bar{\omega}\theta^2}{\zeta}. \tag{31.25}$$

This quantity increases with an increase in the divergence angle of the light, since the fraction of spontaneous emission retained in the resonator is proportional to the solid angle $\pi\theta^2$ within which the stimulated emission is concentrated.

The contribution of the spontaneous emission is insignificant for a small angular divergence (§ 31.1). However, for a large divergence, which leads to a narrowing of the generation spectrum, the contribution of the spontaneous emission must be considered. Since the peak of the spectral distribution examined in § 31.2 narrows when the angular divergence θ increases, while the contribution of the spontaneous emission to the spectral width increases, an optimum value $\theta = \theta_{min}$ exists which corresponds to the smallest resultant magnitude of the spectral width of the generation. This minimum is reached for those values of θ for which the quantities (31.17) and (31.25) are of an identical order of magnitude.

† Spectral narrowing is discussed in references [118, 137].

From this we obtain

$$\theta_{\min} \sim \left(\zeta \frac{\delta\omega_{\parallel}}{\bar{\omega}}\right)^{3/8} \left(\frac{\lambda}{a}\right)^{1/4}, \tag{31.26}$$

and the smallest value of the resultant spectral broadening is

$$\delta\omega_{\min} \sim \delta\omega_{\parallel} \left(\frac{\bar{\omega}}{\zeta\delta\omega_{\parallel}}\right)^{1/4} \sqrt{\frac{\lambda}{a}}. \tag{31.27}$$

§ 32. The Effect of Kinetic Operation on the Spectral Width of the Generation

If the discrete spectrum of the electromagnetic field is not considered (for example, if the wavelength is caused to vanish), then the spectral width of the generated radiation must decrease with time according to the law [60]†

$$\delta\omega \sim \frac{\bar{\omega}}{\sqrt{v\varkappa_1 t}}, \tag{32.1}$$

where t is the time measured from the instant at which generation begins. Actually, from Eq. (25.12) it follows that the gains at the frequencies ω_0 and ω differ by the amount $\varkappa_1[(\omega - \omega_0)/\bar{\omega}]^2$. Therefore, the ratio of the spectral intensities at the points ω and ω_0 is a quantity of the order of $\exp[-v\,\varkappa_1 t(\omega - \omega_0)^2/\bar{\omega}^2]$ at time t; it is from this that (32.1) is derived.

However, the spectral width cannot always decrease according to (32.1), since a decrease in the number of generated longitudinal modes causes an ever stronger manifestation of that spatial nonuniformity of the population inversion which is connected with the discrete structure of the spectrum, and therefore an ever stronger tendency toward the simultaneous generation of many longitudinal modes (see § 15) develops. The narrowing of the spectrum according to (32.1) continues only until a certain time $\tilde{t}$ which will be estimated below.

If the duration t of the generation is less than the value of $\tilde{t}$, then the spectral width of the generation is determined from Eq. (32.1). In the case of a fairly short duration of the generation the

† The decrease of the spectral width of the generation with time according to the law (32.1) was observed experimentally in the case of nonresonance feedback in which the radiation has a practically continuous spectrum [3].

spectral width for transient operation turns out to be greater than its continuous-wave value considered in §31. Thus, for a short duration of the generation its transient nature leads to an increase of the spectral width compared with the continuous-wave case.

On the other hand, as we shall show below, for a sufficiently long duration of the generation (comparable with $\tilde{t}$), the spectral width of the generation for proper kinetic operation can be less than the continuous-wave value. This is connected with the fact that in the process of establishing continuous-wave operation the spectral width passes through a minimum at times $t \sim \tilde{t}$. Thus, in the case of proper kinetic operation the transient nature of the generation can lead to a narrowing of the spectrum compared with the continuous-wave case.

Let us examine this narrowing of the spectrum using the example of a plane-parallel resonator having an infinite cross section (actually, the treatment is applicable to a resonator having finite dimensions with small angular divergence of the light; see § 31.1).

Let us write the kinetic equation for the number of excited atoms which is analogous to Eq. (25.15) but takes account of the spatial nonuniformity of the population inversion:

$$T\frac{\partial \nu(z, t)}{\partial t} = -\nu - w(z, t)(\nu + 1) + \zeta. \qquad (32.2)$$

Here $\nu = [n(z, t)/n^*] - 1$ is the relative excess above the threshold number of excited atoms; $w = Jv\varkappa_1/N^*$ is the energy of stimulated emission in certain units, which can be distributed over the longitudinal modes:

$$w(z, t) = \sum_m B_m(t)\sin^2\frac{\pi m z}{l} = \frac{1}{2}\sum_m B_m - \frac{1}{2}\sum_m B_m \cos\frac{2\pi m z}{l}. \qquad (32.3)$$

The total energy of the electromagnetic field of the m-th longitudinal mode, referred to a unit volume, is equal to

$$w_m = \frac{B_m}{2}. \qquad (32.4)$$

The variation of this energy with time is described by the equation

$$\dot{w}_m = \frac{\dot{B}_m}{2} = -B_m\frac{v}{l}\int_0^l \varkappa(\omega_m, z)\sin^2\frac{\pi m z}{l}\,dz. \qquad (32.5)$$

(The effective absorption coefficient, averaged over the resonator volume with the weight of the spatial intensity distribution of the m-th mode, appears in the right side of this equation.)

Equations (32.2) and (32.5) form a closed system, but in view of the considerable mathematical difficulties associated with finding its exact solution we shall restrict ourselves to estimating the spectral width.

We shall assume that the deviation of n from the threshold value is small (see Chap. VII), so that

$$\nu \equiv \frac{n - n^*}{n^*} \ll 1. \tag{32.6}$$

Making use of Eqs. (25.12) and (25.13), as well as the substitution (32.6), we represent the effective absorption coefficient in the form

$$\varkappa(\omega_m) = \varkappa_1 \left[-\nu + \frac{(\omega_m - \omega_0)^2}{\overline{\omega}^2} \right] \equiv \varkappa_1(-\nu + \xi_m), \tag{32.7}$$

where we have used the following substitution in order to simplify the notation:

$$\xi_m = \frac{(\omega_m - \omega_0)^2}{\overline{\omega}^2}.$$

Substituting Eq. (32.7) into Eq. (32.5), we express the rate of change of the mode intensities in terms of the Fourier components of the population inversion:

$$\dot{B}_m = v\varkappa_1 B_m(\bar{\nu} - \nu_m - \xi_m), \tag{32.8}$$

where

$$\nu_m = \frac{1}{l}\int_0^l \nu(z)\cos\frac{2\pi m z}{l}\,dz, \qquad \bar{\nu} = \frac{1}{l}\int_0^l \nu(z)\,dz. \tag{32.9}$$

We represent the integrated radiation energy (32.3) in the form

$$w(z, t) = w_0(t) + w_1(z, t), \tag{32.10}$$

where

$$w_0(t) = \frac{1}{2}\sum_m B_m(t), \tag{32.11}$$

$$w_1(z, t) = -\frac{1}{2}\sum_m B_m(t)\cos\frac{2\pi m z}{l}. \tag{32.12}$$

Now we make use of the assumption that the kinetic operation is proper (i.e., the assumption that the amplitudes B_m of all the modes vary with time according to an approximately identical law). From this it follows that for all of the generated modes the amplitudes B_m are of an identical order of magnitude at any instant in time. Assume that $\tilde{m}$ modes are generated simultaneously, where

$$\tilde{m} \gg 1. \tag{32.13}$$

Here the sign-constant sum (32.11) containing m approximately identical terms considerably exceeds the analogous sign-variable sum (32.12); i.e.,

$$w_1(z, t) \ll w_0(t). \tag{32.14}$$

The inequality (32.14), which is the mathematical expression of the regularity of kinetic operation, has a simple physical meaning. If radiation consisting of many modes varies as a whole with time, then the resultant intensity is distributed almost uniformly, so that the relative magnitude of the nonuniformities is small (in particular, this also applies to continuous-wave operation, § 15). Here we again encounter the interrelationship between the spatial uniformity of the system and proper kinetic operation.

Let us substitute (32.10) into (32.2) while neglecting the product of the small quantities w_1 and ν, and let us apply a Fourier transformation to Eq. (32.2). We have

$$T\dot{\nu}_m + \nu_m[1 + w_0(t)] = \frac{B_m}{4}. \tag{32.15}$$

Let us solve Eq. (32.15) for ν_m:

$$\nu_m = \frac{1}{4T}\int_0^t B_m(t')\exp\left\{-\int_{t'}^{t}[1 + w_0(t'')]\frac{dt''}{T}\right\}dt'. \tag{32.16}$$

We shall consider the cases of small and large values of the duration t of the generation separately.

1. $t \gg T/(1+\zeta)$. It is easy to confirm the fact that in this case Eqs. (32.8) and (32.16) have a continuous-wave solution. Actually, assuming $B_m(t) = B_m(\infty)$, we find $\nu_m = B_m/4(w_0 + 1)$ from Eq. (32. (32.16). Substituting this into Eq. (32.8) and assuming $\dot{B}_m = 0$, we obtain the equations for continuous-wave values of B_m; the continuous-wave spectral width (15.11) derives from this equation.

2. $t \ll T/(1+\zeta)$, i.e., the duration of the generation is short compared with the time required to establish continuous-wave operation. A single pulse can serve as an example.

For $t \ll T/(1+\zeta)$ Eq. (32.15) can be simplified and takes the form

$$\nu_m = \frac{1}{4T}\int_0^t B_m(t')\,dt'.$$

Having substituted ν_m into Eq. (32.8), divided by B_m, and integrated with respect to time, we find

$$B_m(t) = \text{const}\exp\left\{-v\varkappa_1\left[\xi_m t + \int_0^t \frac{dt'}{4T}\int_0^{t'} B_m(t'')\,dt'' - \int_0^t \bar{\nu}dt\right]\right\}. \tag{32.17}$$

(The factor in front of the exponential does not depend on m, since at the instant $t = 0$ corresponding to the beginning of generation the modes considered had practically identical amplitudes.)

The intensity distribution over the modes is determined by the first two terms in the exponent of Eq. (32.17), since the third term does not depend on m. For small t the first term predominates (since the second depends quadratically on t), and Eq. (32.17) takes the form

$$B_m = B_{m_0}(t)\,e^{-v\varkappa_1\xi_m t} = B_{m_0}(t)\,(1 - v\varkappa_1\xi_m t + \ldots) \tag{32.18}$$

(the index m_0 corresponds to the maximum of the spectral distribution). Thus, the equation (32.1) is satisfied for small t in accordance with the analysis presented.

For sufficiently large t the second term in the exponent of Eq. (32.17) is comparable with the first in order of magnitude. This means that in view of the nonuniformity of the gain the relationship (32.1) ceases to be valid, and the spectral width begins to increase with time. Therefore, the spectral width passes through a minimum at a certain time $\tilde{t}$. In order to estimate the time $\delta\tilde{\omega}$ and the corresponding minimum value of the spectral width $\tilde{t}$, we substitute (32.18) into the right side of Eq. (32.17):

$$B_m = B_{m_0}\exp\left\{v\varkappa_1\left[-\xi_m t + \frac{\xi_m v\varkappa_1}{4T}\int_0^t (t-t')\,t' B_{m_0}(t')\,dt'\right]\right\}. \tag{32.19}$$

In estimating the second term in the square brackets it is necessary to take account of the fact that the time-averaged value of B_{m_0} is a quantity of the order of the average energy w_0 divided by the number of generated modes $\tilde{m}$; however, the average value of w_0 is equal to ζ. From this we obtain the estimate for Eq. (32.19):

$$B_m(t) \sim B_{m_0} \exp\left\{-v\varkappa_1\xi_m t\left(1 - \frac{v\varkappa_1 t^2\zeta}{24T\tilde{m}}\right)\right\}. \qquad (32.20)$$

Therefore,

$$\tilde{t} \sim \sqrt{\frac{T\tilde{m}}{v\varkappa_1\zeta}}, \quad \widetilde{\delta\omega} \sim \left(\frac{\overline{\omega}^4\delta\omega_0\zeta}{v\varkappa_1 T}\right)^{1/5}, \quad \tilde{m} \sim \left(\frac{\overline{\omega}^4\zeta}{\delta\omega_0^4 v\varkappa_1 T}\right)^{1/5}. \qquad (32.21)$$

Here $\delta\omega_0 = \pi v/l$ is the spectral interval between neighboring longitudinal modes; $\tilde{m} = \delta\tilde{\omega}_0/\delta\omega_0$ is the number of generated modes.

This treatment can be generalized for the case of a large number of transverse modes $Z_\perp$ satisfying the criterion (31.23). Then [108, 113, 114]

$$\tilde{t} \sim 10\left[\frac{T^2\tau^3 Z_\perp^2}{\zeta^3}\left(\frac{\overline{\omega}l}{\pi\overline{\overline{v}}}\right)^2\right]^{1/5}. \qquad (32.22)$$

From the analysis which has been carried out it follows that the ratio between the amplitudes of the modes changes substantially during the time $\tilde{t}$. In the case of a resonator with a small angular divergence we have $\tilde{m} \sim 20$ under ordinary experimental conditions in accordance with (32.21), and $\tilde{t}$ has an order of magnitude coinciding with the period of the intensity oscillations. Therefore, the radiation does not behave as a single whole, and the kinetic equation is inapplicable.

However, in the case of a large angular divergence satisfying the inequality (31.23) $\tilde{t}$ substantially exceeds the period of the oscillations; therefore, the radiation varies with time practically as a single whole, so that the intensity oscillations are regular and are described by the kinetic equations. As $Z_\perp$ increases, the regular kinetic operation becomes more stable relative to various perturbations.

For operation with a single pulse or with isolated peaks (Chaps. X–XI) $\tilde{t}$ exceeds the duration of the generation regardless of the angular divergence, so that the spectral width is determined by Eq. (32.1). The maximum narrowing of the spectrum is achieved in the case of a single pulse which is obtained by passive Q-modulation, since for passive modulation the time required for the development of a single pulse is greater than it is for active modulation. This is in agreement with experimental data [16, 27].

Conclusions

1. In the case of a small angular divergence the magnitude of the spectral interval covered by the generated modes is practically no different from the value corresponding to a plane-parallel resonator for $a \to \infty$, $\theta \to 0$ (§ 15).

2. A sufficiently large angular divergence of the light ($\theta \gg \sqrt{\delta\omega_{\parallel}/\omega_0}$) leads to a substantial narrowing of the spectrum of the generated radiation.

3. Proper kinetic operation facilitates the reduction of the spectral width of the generation. In the case of a single pulse the spectral width is less for passive Q modulation than it is for active modulation.

4. In the case of a large angular divergence satisfying the criterion (31.23) the value of $\tilde{t}$ exceeds the value given by (32.21) by approximately a factor of $p^{-4/5}Z_{\perp}^{2/5}$ and reaches a value equal to the duration of the pumping pulse. This provides for regularity of kinetic operation and applicability of the kinetic equation.

Chapter IX

THE THRESHOLD PHENOMENA RELATED TO MICROINHOMOGENEITY OF THE ACTIVE MEDIUM

In the preceding chapters it was assumed that all of the luminescence centers were identical, so that the spatial nonuniformity of the active medium could be caused solely by nonuniformity of the population inversion. In the present chapter we examine the case in which the inhomogeneity of the active medium is connected with the presence of active optical centers of various types; here the distribution of the centers among these types is assumed to be random. At the same time the active medium is assumed to be macroscopically uniform, so that the presence of the centers of different types does not affect these conventional macroscopic properties and can be manifested solely in experiments with stimulated emission. The most essential consequence of the microinhomogeneity investigated is the existence of threshold phenomena which are connected with the fact that the microinhomogeneity can be manifested only in the presence of a sufficiently high pump intensity.

We shall consider both the inhomogeneity connected with the random orientation of the dipole moments of individual luminescence centers and the inhomogeneity connected with the random shift of the luminescence bands belonging to individual centers. In the first case the microinhomogeneity of the active medium affects the polarization of the radiation, while in the second it affects its spectral composition.

Let us begin by considering the first case. Assume that the radiation losses in the resonator depend on the direction of the polarization.

The dependence of the light losses on the direction of the polarization can easily be stipulated experimentally by means of transparent plates with plane-parallel faces mounted in the cavity at an angle with respect to the optic axis [25]. If this angle is equal to the Brewster angle, then the anisotropy of the light losses reaches a maximum. By varying the number of plates and their alignment it is possible to vary the anisotropy of the losses continuously within any limits, beginning at zero.

The smallest loss accompanying reflection from the plates is undergone by light polarized in the plane passing through the optic axis and the normal to the plates; this plane defines the predominant direction of polarization. The direction of the electric vector of the second, weaker polarization component forms a right angle with this plane. Therefore, both directions of polarization can be assumed to be stipulated in advance.

The ratio between the intensities of the polarization components naturally depends on the ratio between their losses. Let us write the ratio between the light-energy losses of the weaker polarization component and the losses of the fundamental component in the form $1 + \gamma$; by definition, $\gamma \geq 0$. For $\gamma = 0$ the light is unpolarized, and for $\gamma \to \infty$ it is completely polarized. However, the degree of polarization depends not only on the anisotropy of the losses, which is stipulated by the parameter γ, but also on the parameter ξ which stipulates the relative excess pump power above the generation threshold. This latter dependence is connected with the nonlinear interaction between the electromagnetic fields which are polarized in different directions. Let us dwell on this in greater detail.

We encountered nonlinear interaction between different modes of the generated radiation earlier (for example, in Chap. V where we considered the simultaneous generation of many longitudinal modes in connection with their nonlinear interaction). The physical meaning of this interaction consists in the fact that it is predominantly excited atoms located near the maxima of the mode intensity which participate in the generation of a given longitudinal mode; the atoms located near the nodes of the mode hardly interact at all with its electromagnetic field. For sufficiently strong pumping one mode is originally generated and a large number of excited atoms develops at its nodes; this number is sufficient for the generation of other modes for which the spatial distribution of the intensity overlaps the distribution of the excess excited atoms.

An analogous mechanism can lead to the simultaneous generation of not only modes but generally different types of radiation.
The generation of each type is connected with a portion of the excited atoms that is characterized by a definite attribute rather than
with all of the excited atoms. In particular, this applies to electromagnetic fields having different directions of polarization. In this
case the generation of each polarization component is connected
predominantly with those excited atoms whose dipole moments form
small angles with the direction of polarization.

Assume that only the one polarization component which
undergoes the smallest light loss is generated. Its electromagnetic
field is emitted predominantly by those excited atoms whose dipole
moments are directed at small angles with respect to the electric
vector; at the same time the pumping (which we shall assume to be
isotropic) excites atoms regardless of the direction of the dipole
moments. Therefore, as the pumping increases, the excited atoms
having dipole moments at right angles to the main direction of the
polarization begin to predominate. Correspondingly, the gain
is greater for the second polarization direction than for the main
direction, and the difference increases as the parameter ζ describing the pumping increases. For a sufficiently large value of ζ
equal to ζ^* the difference between the gains for the two polarization
directions compensates for the difference in the magnitude of the
light-energy losses, and for $\zeta > \zeta^*$ both polarization components
are generated simultaneously. However, if $\zeta < \zeta^*$, then the weaker
component is hardly generated at all.[†] at all.[†]

We shall call the quantity ζ^* the generation threshold for
the weaker component, or, more briefly, the second generation
threshold.[‡]

The magnitude of the second threshold depends not only on
the difference between the losses for the polarization components,
but also on the probability of nonradiative excitation-energy transfer between luminescence centers. If migration of excitations be-

---

[†] The intensity and spectral composition of the weaker components were investigated
in [60, 65] for $\zeta < \zeta^*$.

[‡] The second generation threshold exists only for $\gamma \gg 1/\nu\varkappa_1 t$; for $\gamma < 1/\nu\varkappa_1 t$ the
difference between the intensities of the polarization components and their thres-
hold vanishes [60]. Below we shall assume that $\gamma \gg 1/\nu\varkappa_1 t$ (t is the duration of
the generation).

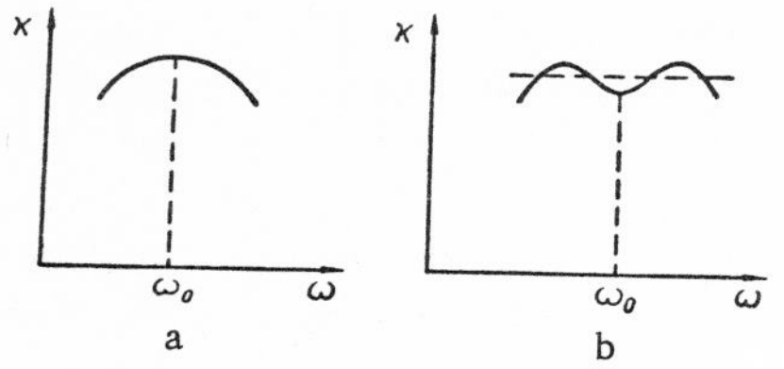

Fig. 34. Diagram of the shape of the amplification band: a) the pump intensity does not reach the spectral broadening threshold; b) the pump intensity exceeds this threshold.

tween luminescence centers occurs (§ 3), then it leads to a certain smoothing of the dependence of the number of excited atoms on the orientation of the dipole moment and thus to an increase of the second threshold and a decrease of the intensity of the weaker polarization component.

A similar threshold phenomenon can also occur with respect to the spectral width of the generations, if the active medium is inhomogeneous due to a random shift of the luminescence bands of individual centers (it is the practice to say that in this case the observed luminescence band is inhomogeneously broadened).

For a small excess above the generation threshold, for which the intensity of the generated radiation is low and its deexcitation action is insignificant, the stimulated emission cannot noticeably deform the amplification band. The latter has a maximum at the center frequency ω_0 (Fig. 34a), as it does in the absence of generation, so that the inhomogeneous broadening of the luminescence does not affect the spectral composition of the generation.

As the pumping increases, the intensity of the stimulated emission, and therefore of its deexciting action, similarly increases. Under these conditions it is the active centers, whose luminescence maximum is close to the center frequency ω_0 near which generation occurs, which are predominantly deexcited; however, the pumping excites all of the active centers to an approximately identical degree. Therefore, for a sufficiently strong pumping the amplification bands can alter its shape, as shown in Fig. 34b; under these conditions the generation includes an entire band of frequencies, so that the inhomogeneous broadening of the luminescence band entails the broadening of the generated line. This phenomenon has a threshold character; it occurs only for a sufficiently large relative excess ζ above the generation threshold. The corresponding critical value $\zeta = \bar{\zeta}$ shall be called the spectral-broadening threshold. The quantity $\bar{\zeta}$ increases with an increase in the number of excitation migrations s, just as the threshold of the weaker polarization components does.

§33. Spectral Width of the Radiation from Lasers Having an Inhomogeneously Broadened Luminescence Line. The Basic Equations

In order to investigate the spectral properties of a laser having an inhomogeneously broadened luminescence line† it is necessary to eliminate the spatial nonuniformity of the population of the inversion and its effect on the spectral composition of the generation (see Chap. V). This can be achieved, for example, by using a moving active medium (Chap. V) or a resonator having a large angular divergence of the light (Chap. VIII).

In accordance with the above we shall not consider the phase relationships and shall assume that the inhomogeneous broadening of the luminescence line is the sole cause of laser line broadening.

The observed luminescence band consists of elementary bands corresponding to individual active centers; these elementary bands differ only in the position of their maximum but not in their shape. Considering a four-level diagram in order to be specific, we write the gain as a function of the frequency:

$$\tilde{\varkappa}(\omega) = Bn \int P(\omega - \omega') f(\omega')\, d\omega'. \tag{33.1}$$

Here n is the total number of excited active centers per unit volume; $f(\omega)$ is their distribution function with respect to the position of the luminescence maximum (normalized per unit area); $P(\omega)$ is the shape of the elementary luminescence band; B is a certain constant which is proportional to the oscillator strength of the active centers.

Assuming for simplicity that the generation is continuous, we write the equation for $f(\omega)$:

$$n\frac{\partial f}{\partial t} = 0 = -\frac{nf}{T} - \frac{Bvnf}{\hbar\omega_0} J \int P(\omega - \omega') \varrho(\omega')\, d\omega' + \\ + \frac{Nf_0(\omega)}{\hbar\omega_0} + \frac{snf_0}{T} - \frac{snf}{T}. \tag{33.2}$$

Here J is the volume energy density of the stimulated emission;

† An example of active media having an inhomogeneously broadened luminescence line can be found in various glasses and crystals which have been activated with rare-earth elements. The cause of the inhomogeneous broadening is the random shift of the luminescence line of the rare-earth ion, which is connected with the irregularity of the fields created by the surrounding atoms.

$\rho(\omega)$ is its spectral distribution; $f_0(\omega)$ is the distribution function of the unexcited centers with respect to the position of the luminescence maximum; s is the probability of excitation migrations between active centers during a time T. All of the distribution functions which are used have been normalized per unit area.

Equation (33.2) is directly obvious: the first two terms on the right side describe the spontaneous and stimulated emission from excited atoms; the third term describes the excitation of the atoms by the pumping, and the last two terms describe the migration of excitations. We have taken account of the fact that the probability of stimulated emission from centers of a stipulated type is proportional to the integral of the overlap between their elementary amplification bands and the spectral distribution of the generated light. However, the energy of the pumping and the migrating excitations is distributed among the different types of centers in proportion to their number in the ground state (this number is assumed to be large enough so as to be independent of the pump intensity).

Solving Eq. (33.2) for f and substituting into (33.1), we express the gain in terms of the spectral density of the stimulated emission:

$$\tilde{\varkappa}(\omega) = \frac{Bn}{1+s}\left(\frac{NT}{\hbar\omega_0 n}+s\right)\int \frac{P(\omega-\omega')\, f_0(\omega')\, d\omega'}{1+(u^2-1)\int P(\omega'-\omega'')\,\varrho(\omega'')\, d\omega''}\cdot \qquad (33.3)$$

Here we have introduced the quantity

$$u^2 = 1 + \frac{BTv}{\hbar\omega_0(1+s)} J, \quad u \gg 1, \qquad (33.4)$$

which is the measure of the stimulated-emission energy.

In order to simplify the calculations we shall assume that the functions f_0 and P are Lorentz functions†

$$P(\omega) = \frac{\overline{\omega}^2}{\omega^2+\overline{\omega}^2}, \quad P(0) = 1; \qquad (33.5)$$

$$f_0(\omega) = \frac{q\overline{\omega}}{\pi}\cdot\frac{1}{(\omega-\omega_0)^2+\overline{\omega}^2 q^2}, \qquad \int f_0 d\omega = 1 \qquad (33.6)$$

Here $2\,\overline{\omega}$ is the half-width of the elementary luminescence band; q is the ratio between the spread of the elementary bands with res-

† Usually, f_0 is a Gaussian function, but this cannot have a substantial effect on the result.

pect to their maximum position and the half-width of the elementary band. The parameter q (which is important further on) characterizes the degree of inhomogeneity of the luminescence band.

Below we shall assume that the spectral width of the generation is considerably narrower than the elementary luminescence band, i.e.,

$$\delta\omega \ll \bar{\omega}. \tag{33.7}$$

Then it is possible to assume that $\rho(\omega)$ is a δ-function. Substituting (33.5) and (33.6) into Eq. (33.3) and carrying out elementary integration in the latter, we have

$$\tilde{\varkappa}(\omega) = \frac{Bn}{1+s}\left(\frac{NT}{\hbar\omega_0 n} + s\right)\frac{q}{u^2 - q^2}\left\{\frac{1-q}{q\left[1 + \frac{(\omega-\omega_0)^2}{\bar{\omega}^2 (q+1)^2}\right]} - \frac{1-u}{u\left[1 + \frac{(\omega-\omega_0)^2}{\bar{\omega}^2 (u+1)^2}\right]}\right\}. \tag{33.8}$$

Making use of the condition (33.7), we expand this expression in powers of $\omega - \omega_0$:

$$\tilde{\varkappa}(\omega) = \frac{Bnq}{1+s}\left(\frac{NT}{\hbar\omega_0 n} + s\right)\left\{\frac{1}{qu(q+u)} + \frac{F_2(u) - F_2(q)}{u^2 - q^2}\left(\frac{\omega-\omega_0}{\bar{\omega}}\right)^2 + \frac{F_4(q) - F_4(u)}{u^2 - q^2}\left(\frac{\omega-\omega_0}{\bar{\omega}}\right)^4 + \ldots\right\}. \tag{33.9}$$

Here we have introduced the substitutions

$$F_2(u) = \frac{1-u}{u(u+1)^2}, \quad F_4(u) = \frac{1-u}{u(u+1)^4}. \tag{33.10}$$

§ 34. The Spectral-Broadening Threshold [115]

Let us fix the parameter q and vary the quantity u (i.e., the stimulated-emission energy). Assume first that $q < 1$; then it is not difficult to see that the coefficient of the quadratic term in the expansion (33.9) is negative for any u within an allowed interval $u \geq 1$, so that the amplification band has the form shown schematically in Fig. 34a. In this case generation takes place only at the frequency ω_0, and inhomogeneous broadening of the luminescence line does not lead to broadening of the generated line.

Assume now that $q > 1$. Then at the point $u = \bar{u}(q)$ satisfying the equation

$$F_2(\bar{u}) = F_2(q), \quad \bar{u} \neq q, \tag{34.1}$$

the coefficient of the quadratic term in the expansion (33.9) vanishes, while for $u > \bar{u}(q)$ this coefficient becomes positive; however, the coefficient of the fourth-power term is negative. Therefore, for $u > \bar{u}(q)$ the gain reaches a maximum at two symmetrical points (Fig. 34b), rather than at the point ω_0. Therefore, generation cannot occur just at the frequency ω_0, but covers a spectral interval whose order of magnitude coincides with the interval between maxima.

From what has been said it is clear that the quantity $u(q)$ is the spectral-broadening threshold; actually, inhomogeneous broadening of the luminescence line leads to broadening of the generation line only for $u > \bar{u}(q)$.

The variable u is connected with the radiation intensity J by Eq. (33.4); in turn, J can be expressed in terms of the relative excess ζ above the threshold pump power. Thus, we find the critical value $\zeta = \bar{\zeta}$ corresponding to the spectral-broadening threshold:

$$\bar{\zeta} = g_1(q) + sg_2(q), \tag{34.2}$$

where we have introduced the functions

$$g_1(q) = \frac{\bar{u}(\bar{u} + q) - 1 - q}{q + 1}, \quad g_2(q) = \frac{\bar{u}^2 - 1}{q + 1}. \tag{34.3}$$

Let us provide the asymptotic expansion for these functions:

$$g_1 = \frac{2}{q - 1} + \ldots, \quad g_2 = \frac{2}{q - 1} + \ldots \quad \text{for} \quad q - 1 \ll 1, \tag{34.4}$$

$$g_1 = \frac{4}{q^2}\left(1 - \frac{2}{q} + \ldots\right), \quad g_2 = \frac{8}{q^3}\left(1 - \frac{4}{q} + \ldots\right) \text{for } q \gg 1 \tag{34.5}$$

The graphs of the functions g_1 and g_2 are shown in Fig. 35. From the figure and from Eqs. (34.4), (34.5) it is obvious that these functions decrease monotonically with increasing q; the spectral-broadening threshold decreases accordingly (for stipulated s).

A physical explanation of the results obtained is not difficult. The existence of a spectral-broadening threshold is connected with

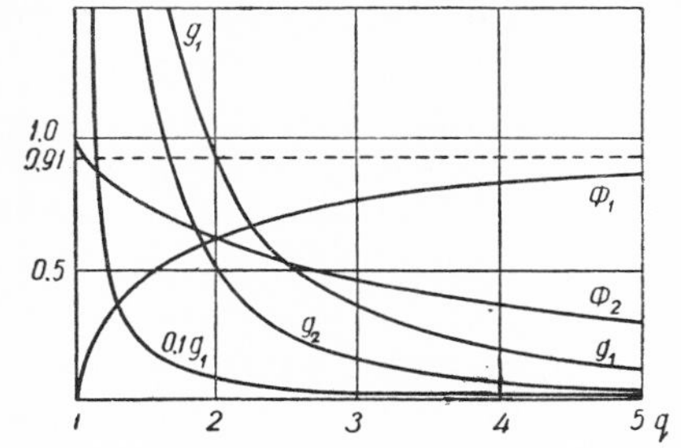

Fig. 35. Graphs of the functions that determine the spectral generation characteristics connected with the spectral broadening threshold.

the fact that the generated radiation predominantly deexcites those excited atoms whose gain maximum ω_{max} is close to the generation frequency ω_0. For a sufficiently high generation intensity, which is proportional to the pump power above-threshold, this leads to the deformation of the resultant amplification band shown in Fig. 34b. As q increases, the spectral overlap between the generation line and the amplification band of active centers having $\omega_{max} \neq \omega_0$ becomes less. Therefore, as q increases, $\bar{\xi}$ decreases (i.e., the pump power above-threshold required for a substantial deformation of the amplification band decreases).

Because of migration, the excitations are redistributed among active centers having different ω_{max} in proportion to their number in the ground state (i.e., in proportion to the function $f_0(\omega)$). This produces a tendency toward restoration of the initial shape of the amplification band (Fig. 34a), which leads to an increase of the spectral-broadening threshold $\bar{\xi}$. Therefore, $\bar{\xi}$ increases as s increases.

Let us examine the spectral width of the generation for $\xi > \bar{\xi}$; in this we shall assume for simplicity that the excess of the spectral-broadening threshold is small enough so that condition (33.7) is satisfied. Then Eq. (33.9) for $\tilde{\varkappa}$ is valid; this equation has a maximum at two frequencies which are symmetrical with respect to ω_0 (Fig. 34b). Such a shape of the amplification band cannot, strictly speaking, exist during the process of continuous generation, since within the spectral interval occupied by the generated light the gain must compensate the light losses and therefore cannot depend on frequency. The dependence of the gain on frequency shown in Fig. 34b derives from the fact that in Eq. (33.3) we neglected the width of the spectral distribution of the light in comparison with the widths of the elementary luminescence bands. Consideration of the finite spectral width would lead to an insignificant deformation of the amplification band while conserving its area (i.e., while conserving the total number of excited atoms). As a result, a flat region would appear on the $\tilde{\varkappa}(\omega)$ curve (this region is shown in Fig. 34b by the horizontal dashes); the width of this region is equal to the spectral width of the generation, and the area under it coincides with the area under the initial curve. Taking this concept into account and making use of Eq. (33.9), we easily obtain the estimates for the spectral width of the stimulated emission for a slight excess above the spectral-broadening threshold:

$$\frac{\delta\omega}{\Delta\omega} \cong \Phi_1(q)\sqrt{\frac{\zeta-\bar{\zeta}}{1+s\Phi_2(q)}}. \tag{34.6}$$

Here $\Delta\omega$ is the half-width of the observed luminescence band,

$$\Phi_1 = \sqrt{\frac{5}{6}}\sqrt{\frac{F_2'(\bar{u})}{(1+q)(2\bar{u}+q)[F_4(\bar{u})-F_4(q)]}}, \quad \Phi_2 = \frac{2\bar{u}}{2\bar{u}+q}.$$

The graphs of the functions Φ_1 and Φ_2 are shown in Fig. 35; their asymptotic expansions have the form

$$\Phi_1 = \sqrt{\frac{5}{18}(q-1)} + \ldots, \quad \Phi_2 = 1 - \frac{\sqrt{q-1}}{4} + \ldots, \quad q-1 \ll 1, \tag{34.7}$$

$$\Phi_1 = \sqrt{\frac{5}{6}} - o\left(\frac{1}{q^2}\right), \quad \Phi_2 = \frac{2}{q}\left(1-\frac{2}{q}+\ldots\right), \quad q \gg 1. \tag{34.8}$$

Thus, for $\zeta > \bar{\zeta}$ the spectral width of the generation increases in proportion to the square root of the magnitude of the excess above the spectral-broadening threshold as the pumping increases. The coefficient of this square root increases as the parameter q characterizing the inhomogeneity of the luminescence band increases, and it decreases as the number of excitation migrations s increases.

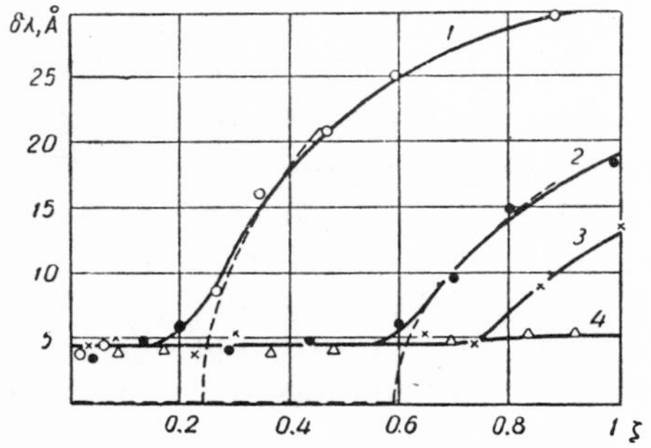

Fig. 36. The dependence of the spectral width of the radiation from a neodymium-glass laser on the relative excess above the generation threshold at various temperatures: the solid lines represent the experimental data; the dashed lines represent the calculated results.

From what has been said it follows that there is a substantial dependence of the spectral width on the pump intensity; this dependence is considerably stronger than it is in the case of a homogeneously broadened luminescence line. This fact was noted in [29], the substantial effect of the migration of excitations on the spectral width of the generation was also emphasized in [29, 102].

The theory presented above has been confirmed experimentally [115]. Figure 36 shows the experimental dependence of the spectral width of the generation on the relative excess above the threshold pump power at various temperatures [1) 20°; 2) 60°; 3) 80°; 4) 90°C] for a neodymium–glass laser having a large angular divergence. The dashed curves represent the results calculated according to the equations of the preceding sections, and the parameters q and s were chosen in such a way as to obtain the best agreement with the experimental points.

From the figure it is evident that the theory developed above provides a quantitative description of the experimental relationships. A certain discrepancy between theory and experiment develops in the range $\xi < \bar{\xi}$, where the experimentally observed spectral width has a nonzero value of about 5 Å which is practically independent of the pumping. This is connected with factors having no relation to the inhomogeneous broadening of the luminescence line, which are not considered by the theory (in particular, it is connected with the dependence of the position of the luminescence maximum on the temperature which varies during the generation process, and with the macroscopic nonuniformity of the active medium).

From a comparison of the theory with experiment we obtain the following values for the parameters of the active medium: q = 13, s = 90 at 20°C, and q = 12, s = 190 at 60°C.†

§ 35. The Equation for the Intensity of the Polarization Components

Let us carry out the calculations for the simplest case in which the active sample is a uniaxial crystal whose axis coincides with the optic axis of the laser, and the two equivalent crystallo-

† The strong temperature dependence of the migration is possibly connected with the fact that a large q corresponds to a weak overlap of the elementary bands of the individual centers. The small overlap integral is sensitive even to a weak temperature broadening of the elementary bands.

graphic directions form a right angle with one another and with the optic axis. We shall assume that the dipole moments of the luminescence centers can be oriented with equal probability along the equivalent crystallographic axes. Such a crystal is optically isotropic in the plane perpendicular to the optic axis.

The case of an isotropic active medium is somewhat more complex mathematically; therefore we shall omit their derivation and limit ourselves to a presentation of the basic results.

Assume that the direction of the polarization forms an angle φ with one of the two equivalent crystallographic directions; in this it can be assumed that $\varphi \leq \pi/4$ without loss of generality. Assume that a unit volume contains n excited atoms (luminescence centers) having a dipole moment oriented at the angle φ with respect to the principal direction of the polarization, and n' excited atoms whose dipole moment forms the angle $\pi/2 - \varphi$ with this direction (Fig. 37).

Generalizing Eq. (9.4) while considering the orientation of the dipole moments relative to the electric vector of the light wave, we write the effective absorption coefficient for the principal polarization component:

$$\varkappa(\omega_0) = \varkappa_1 - \frac{2\varkappa_1}{n^*}(n\cos^2\varphi + n'\sin^2\varphi) \qquad (35.1)$$

while for the second, weaker polarization component we write

$$\tilde{\varkappa}(\omega_0) = \varkappa_1(1+\gamma) - \frac{2\varkappa_1}{n^*}(n\sin^2\varphi + n'\cos^2\varphi). \qquad (35.2)$$

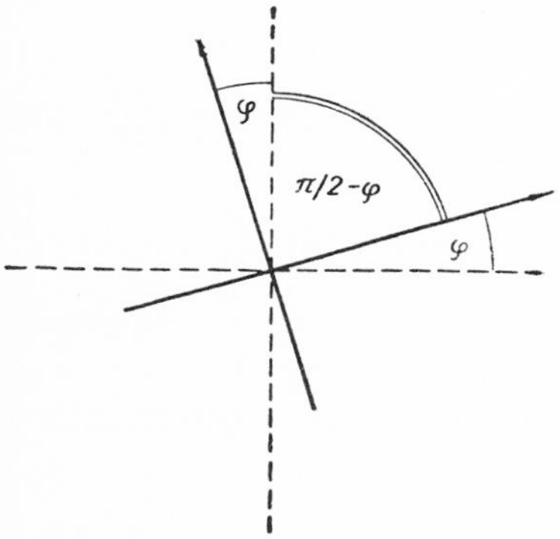

Fig. 37. Orientation of the crystallographic axes (the dashed lines) relative to the polarization direction (the arrows).

Here $\varkappa_1$ are the losses of the principal component, referred to a unit length; $\varkappa_1(1 + \gamma)$ represents the same quantity for the weaker component; n^* is the threshold number of excited atoms per unit volume. For an infinitely small excess above the principal generation threshold, when it is possible to neglect the deexciting action of the generated radiation, we have $n = n' = n/2$ for isotropic pumping, and Eq. (35.1) or (35.2) takes the conventional form (9.4).

We also write the equations for n and n':

$$\frac{dn}{dt} = \frac{N}{2\hbar\omega_0} - \frac{n}{T} - \mathscr{E}n\cos^2\varphi - \mathscr{E}'n\sin^2\varphi + \frac{s}{T}(n' - n), \qquad (35.3)$$

$$\frac{dn'}{dt} = \frac{N}{2\hbar\omega_0} - \frac{n'}{T} - \mathscr{E}n'\sin^2\varphi - \mathscr{E}'n'\cos^2\varphi + \frac{s}{T}(n - n'). \qquad (35.4)$$

Here T is the spontaneous-emission time; $\mathscr{E}$ is the energy of the principal polarization component, expressed in certain units; $\mathscr{E}'$ is the same quantity for the weaker polarization component; s is the probability of excitation migration between atoms having different orientations of their dipole moment in the time T; N is the pump power absorbed per unit volume (it is assumed that the pump intensity is distributed evenly among atoms with both orientations of the dipole moment). The first term in the right side of Eqs. (35.3), (35.4) describes the excitation of the atoms by the pumping; the second describes their spontaneous deexcitation; the third describes their stimulated deexcitation due to the action of the principal polarization component; the fourth describes their deexcitation due to the action of the weaker polarization component; finally, the last term describes the nonradiative excitation-energy transfer between atoms having different orientations of their dipole moments (s/T is the transfer probability per unit time).

Equations (35.1)-(35.4) can be used to find the second generation threshold.

§36. The Second Generation Threshold

Assume that the pump intensity exceeds the principal generation threshold but does not reach the second threshold. Then $\tilde{\varkappa}(\omega_0) > 0$, and the second (weaker) polarization component is practically not generated, so that

$$\mathscr{E}' \ll \mathscr{E}. \qquad (36.1)$$

However, for the principal component the generation condition is satisfied:

$$\varkappa(\omega_0) = 0, \tag{36.2}$$

which means that the radiation losses are compensated by the amplification. For continuous-wave operation, as was shown in [60], condition (36.2) is satisfied with high accuracy during the process of actual generation.

For operation with relaxation oscillations of the intensity (Chap. VII) $\varkappa(\omega_0)$ undergoes small oscillations near zero, but the value of $\varkappa(\omega_0)$ averaged over a period of the oscillations is zero. This derives from the fact that the intensity varies with time in proportion to the quantity $\exp[-v\int\varkappa(\omega_0)dt]$; on the other hand, the shift during a period returns the intensity to its previous value, so that

$$\overline{\varkappa(\omega_0)} = \frac{1}{\Delta t}\int_t^{t+\Delta t}\varkappa(\omega_0, t')\,dt' = 0$$

(Δt is the period). Thus, if in Eq. (35.1) we define n and n' as the values of these quantities averaged over a period, then the left side of this equation vanishes not only for continuous-wave operation but also for operation with relaxation oscillations. This cannot be said of Eq. (35.2), since by assumption the generation condition is not satisfied for the weaker polarization component.)

For a similar reason the left side of Eqs. (35.3), (35.4) vanishes not only for continuous-wave operation but also for operation with relaxation oscillations if n, n', $\mathscr{E}$, and $\mathscr{E}'$ are defined as the values of these quantities averaged over a period of the oscillations.†

Thus, we have obtained three equations with three unknowns n, n', $\mathscr{E}$ ($\mathscr{E}'$ drops out of these equations in accordance with the condition (36.1)). Determining n and n' from these equations and substituting these quantities into (35.2), we find $\tilde{\varkappa}(\omega_0)$, the effective absorption coefficient for the weaker polarization component. The second generation threshold is reached when this quantity vanishes;

† The product $\mathscr{E}_n$ is averaged over the period as follows: $\overline{\mathscr{E}n} = \overline{\mathscr{E}}\overline{n} + \overline{(n-\overline{n})\mathscr{E}}$. The second term vanishes, since $\mathscr{E}$ is an even function of the time measured from the instant of maximum activity, and $n - \overline{n} = \text{const}\ \varkappa(\omega_0)$ is an odd function (see Chap. VII).

under these conditions, by definition $\zeta = \zeta^*$. Finding ζ from the conditions $\tilde{\varkappa}(\omega_0) = 0$, we finally have [54]

$$\zeta^* = \gamma \frac{1 + \cos^2 2\varphi + 2s - 0.5\gamma \sin^2 2\varphi}{2\cos^2 2\varphi - \gamma \sin^2 2\varphi}. \tag{36.3}$$

For $\varphi = 0$ the second threshold reaches the minimum value equal to

$$\zeta^*_{\min} = \gamma(1 + s). \tag{36.4}$$

If φ is increased continuously, beginning at zero, then ζ^* increases monotonically and goes to infinity at the point

$$\varphi_0 = \frac{1}{2} \arctan \sqrt{\frac{\gamma}{2}} \tag{36.5}$$

Physically, this is connected with the fact that for $\varphi = \pi/4$ the difference between the two possible orientations of the dipole moment actually vanishes, since the latter is oriented at an angle $\pi/4$ with respect to both polarization directions in any case (Fig. 33). Therefore, for $\varphi = \pi/4$ it becomes impossible to compensate the additional losses of the weaker polarization component, which are stipulated by the parameter γ, at the expense of the nonuniform distribution of the excited atoms with respect to the orientations of the dipole moment. For values of φ close to $\pi/4$ such compensation is still impossible (i.e., the second threshold goes to infinity). As γ increases, the range of values of φ in which the second threshold ζ^* has a finite value naturally decreases.

The second generation threshold (36.3) increases as the parameters γ and φ increase, and also as the probability of excitation migration s/T increases. The dependence of ζ^* on s is connected with the fact that migration of excitation leads to smoothing of the nonuniform distribution of the excited atoms with respect to the directions of the dipole moment.

In the case of an isotropic active medium ζ^* depends only on γ and s. The general form of this dependence is rather complex, but for $\gamma \ll 1$ the expression for ζ^* takes the simple form

$$\zeta^* = \gamma\left(\frac{5}{6}s + \frac{3}{2}\right). \tag{36.6}$$

The dependence of the second threshold on the parameter γ in the range of larger values of the latter is shown graphically in Fig. 38.

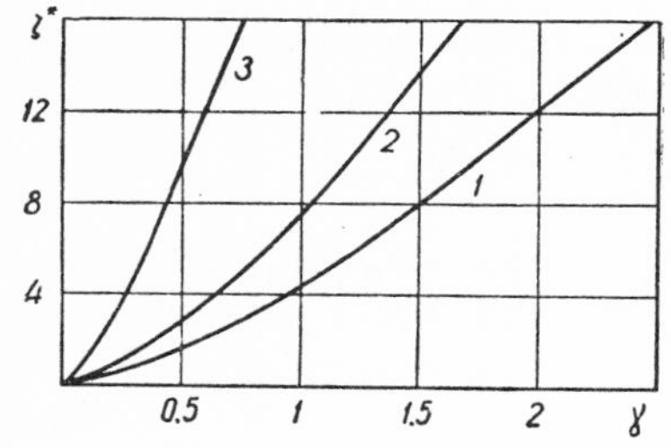

Fig. 38. Dependence of the second generation threshold on the supplementary losses of one of the polarization components for fixed values of the parameter s: 1) s = 0; 2) s = 1; 3) s = 6.

Note that in the case of an isotropic medium, for which the dipole moments can have any direction, the second threshold is finite for any values of γ.

§37. Intensity of the Principal Component under Conditions for Which the Second Component Cannot Be Generated

Let us imagine that due to very large supplementary losses of the second (weaker) polarization component its threshold cannot be reached under the given experimental conditions. The problem arises of the intensity of the principal component under such conditions [64].

The intensity of the principal component can be found from the three equations in n, n', $\mathcal{E}$ that were examined in § 36. Let us give the expression for the volume energy density of the principal component in the limiting cases in which it has a simple form:

$$J = J_0 \frac{1 + 2s}{1 + 2s + \cos^2 2\varphi}$$

for

$$\zeta \ll \frac{1 + s}{\sin^2 2\varphi}, \quad (37.1)$$

and $J = J_0$ for

$$\zeta \gg \frac{1}{\sin^2 2\varphi}. \quad (37.2)$$

Here we have introduced the substitution

$$J_0 = \frac{(N - N^*)}{v\varkappa_1}. \quad (37.3)$$

As was shown in § 11, J_0 is the intensity of the unpolarized light generated by the laser for $\gamma = 0$.

The results obtained have the following physical meaning. For $\varphi = 0$, s = 0 one half of all the excited atoms do not participate in the generation of the principal component, and its intensity is one half the intensity J_0 of the unpolarized light. As φ increases, all of the excited atoms are involved to an ever greater extent in the generation of the principal component, and for the value $\varphi = \pi/4$ at which the difference between the two orientations of the dipole moment vanishes the value of J coincides with the intensity of the unpolarized light. For a similar reason J increases as the parameter s, which is proportional to the probability of migration of the excitations, increases. For $s \gg 1$, regardless of the values of φ and ζ, all of the excited atoms participate to an identical extent in the generation, since the excitation has time to make multiple transfers from atom to atom during its lifetime. Therefore, for $s \gg 1$ we have $J = J_0$. Thus, the ratio J/J_0 increases as φ and s increase. It also increases with an increase in ζ. Actually, as ζ increases, the energy density of the stimulated emission similarly increases, and the involvement of all of the excited atoms in the generation process becomes more efficient. For $\zeta \to \infty$ the ratio J/J_0 tends to unity. In the general case the intensity of the principal component is lower than the intensity of the unpolarized light but greater than one half of this intensity.

A similar situation also occurs in the case of an isotropic active medium; here the excited atoms are distributed continuously with respect to the directions of the dipole moment. In this case

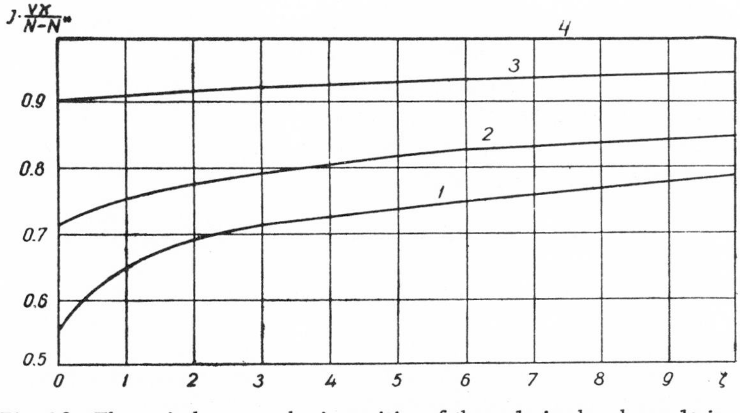

Fig. 39. The ratio between the intensities of the polarized and unpolarized radiation from a laser as a function of the relative excess above the generation threshold for fixed values of the probability s/T of excitation migration: 1) s = 0; 2) s = 1; 3) s = 6; 4) s = ∞.

we have

$$J = J_0 \frac{s+1}{s+\frac{9}{5}} \left[1 + 0.617 \frac{\zeta}{\left(s+\frac{9}{5}\right)^2} + \ldots \right], \quad \zeta \ll s+1, \tag{37.4}$$

$$J = J_0 \left[1 - \frac{0.906}{\sqrt{\zeta(s+1)}} + \left(0.51 + \frac{0.412}{s+1}\right)\frac{1}{\zeta} + \ldots \right], \quad \zeta \gg 1. \tag{37.5}$$

Here $J_0 = (N - N^*)/v\varkappa_1$. The minimum ratio J/J_0 is 5/9, while the maximum ratio is unity. The graph of J/J_0 is shown in Fig. 39.

Thus, if we wish to increase the intensity of one of the polarization components while suppressing the second component, then the effect will increase as s and ζ increase (in the case of a uniaxial crystal the angle φ is also essential).

§ 38. The Degree of Polarization of the Radiation

Let us now examine the case in which the pumping exceeds the second generation threshold, and let us determine the degree of polarization of the radiation [64]. The calculations can be carried out by analogy with §§ 36 and 37, with the sole difference that now we can no longer neglect the energy $\mathscr{E}'$ of the weaker component compared with the energy $\mathscr{E}$ of the principal component. Therefore, instead of three unknowns we have four: n, n', $\mathscr{E}$, and $\mathscr{E}'$. The three equations derived above are not sufficient for finding these unknowns. In order to obtain the fourth equation we make use of the fact that for $\zeta > \zeta^*$ the generation conditions $\varkappa(\omega_0) = 0$ must be satisfied for the second component as well as for the principal component; we also set the right side of Eq. (35.2) equal to zero. From these equations we can find the ratio between the intensities of the polarization components:

$$\frac{\mathscr{E}'}{\mathscr{E}} = \frac{\frac{\zeta}{\gamma}[2\cos^2 2\varphi - \gamma\sin^2 2\varphi] - \cos^2 2\varphi - 1 + \frac{\gamma}{2}\sin^2 2\varphi - 2s}{\frac{\zeta}{\gamma}[2\cos^2 2\varphi + \gamma(\cos^2 2\varphi + 1)] + \left(1 + \frac{\gamma}{2}\right)\sin^2 2\varphi + 2s(1+\gamma)}. \tag{38.1}$$

As we might have expected, this ratio vanishes for $\zeta = \zeta^*$.

The ratio (38.1) reaches its maximum value

$$\left(\frac{\mathscr{E}'}{\mathscr{E}}\right)_{\max} = \frac{1}{1+\gamma} \cdot \frac{\zeta - \gamma - \gamma s}{\zeta + \gamma s} \tag{38.2}$$

at the point $\varphi = 0$. If $\mathcal{E}'/\mathcal{E}$ is continuously increased, beginning with zero, then the ratio $\varphi < \varphi_0 \equiv (1/2)$ arccot $(\gamma/2)^{1/2}$ decreases monotonically and vanishes at a certain point ζ^* at which ζ becomes equal to the stipulated value of ζ (we recall that the second generation threshold ζ^* goes to infinity at the point φ_0). The ratio $\mathcal{E}/\mathcal{E}$ likewise decreases with an increase in γ and s.

We give the formula for $\mathcal{E}'/\mathcal{E}$ in the case of an isotropic active medium. Limiting ourselves to small values of γ for the sake of simplicity, we have

$$\frac{\mathcal{E}'}{\mathcal{E}} = \frac{6\zeta - \gamma(9 + 5s)}{6\zeta + \gamma(3 + 5s)} = \frac{\zeta - \zeta^*}{\zeta + \zeta^* - \gamma} \quad (\gamma \ll 1). \tag{38.3}$$

Let us summarize what has been said above. As s and γ increase (and in the case of a uniaxial crystal as φ approaches $\pi/4$) the characteristics of the laser radiation undergo the following changes:

a) the second threshold ζ^* increases, and under these conditions

$$\zeta^* = \infty \quad \text{for } 0.5 \text{ arccot} \sqrt{\frac{\gamma}{2}} < \varphi < \frac{\pi}{4};$$

b) for a stipulated ζ the degree of polarization increases, and here the radiation is practically completely polarized for $\zeta^* > \zeta$;

c) for a stipulated ζ the intensity of the principal component increases.

The dependence of the intensity of the principal component on ζ has a break at the point $\zeta = \zeta^*$, since for $\zeta < \zeta^*$ it increases more rapidly with an increase in pumping than it does in the range $\zeta > \zeta^*$, where the second component is also generated.

§ 39. Comparison with Experimental Data

An experimental investigation of the radiation polarization in a laser having an isotropic active medium (glass activated with neodymium was reported in [25]). The supplementary losses of one of the polarization components were produced by means of a glass plate rotating about the axis perpendicular to the optic axis of the laser.

The experimental data given in [25] are insufficient for a quantitative comparison with the theory; however, the qualitative relationships presented there are in full agreement with the results of the theory developed above. Experiment shows that for weak pumping which obviously does not reach the second threshold the laser radiation is completely polarized; for a sufficient increase in the pumping, so that it exceeds the second threshold, generation of the weaker polarization component begins. For a further increase in the pumping the radiation becomes unpolarized, and this is in agreement with Eq. (38.3) (this equation is applicable in the present case, since the experiment was carried out for small γ).

Experiment also shows that an increase in the supplementary losses of the weaker component leads to an increase in the degree of polarization for a stipulated pump intensity; under these conditions the degree of polarization increases until the generation of the weaker component ceases altogether. Obviously, the weaker component ceases to be generated at the instant at which the second threshold exceeds the stipulated excess ζ above the threshold power.

Although the treatment in this section was carried out for a four-level diagram, the results are also applicable to a three-level diagram if the substitutions (9.3) are considered.

Conclusions

1. In the case of a homogeneously broadened luminescence line (for a uniform spatial distribution of the population inversion) a spectral-broadening threshold exists. If the pump intensity does not reach this threshold, then inhomogeneous broadening of the luminescence line does not lead to a broadening of the generated line. The spectral-broadening threshold increases as the inhomogeneity of the luminescence line decreases and as the probability of excitation migration increases.

2. Similarly, if the light losses in the cavity depend on the direction of the polarization, then the two polarization components correspond to different generation thresholds; the interval between these thresholds increases as the difference between the losses and the probability of excitation migration increase.

3. These threshold phenomena can be used to determine the probability of excitation migration between active centers.

4. The intensity of the principal polarization component can be increased by suppressing the second component. The maximum

effect – a doubling of the intensity – can be achieved either for very powerful pumping or for a sufficiently high probability of excitation migration.

Chapter X

ACTIVE Q MODULATION †

For certain laser applications it is required to obtain generation in the form of one intensity peak having a short duration. Such a peak is called a single pulse to distinguish it from the train of peaks which accompany relaxation oscillations of the intensity (Chap. VII). A resonator with a rapidly switched Q is used to produce a single pulse.

§ 40. Production of a Single Pulse by Means of Rapid Q Switching‡

In order to produce a single pulse an optical shutter which can be opened rapidly is placed in the cavity. Pumping of the active medium is accomplished with the optical shutter closed, which makes generation impossible. During the pumping process the number n of excited atoms increases to a certain value $\bar{n}$ which exceeds the threshold value $\underline{n}$ corresponding to an open shutter. Then the optical shutter is opened rapidly, and intense generation begins during which the excess number of excited atoms $\bar{n} - \underline{n}$ is deexcited.

In order to estimate the duration of the generation we note that the excess $\bar{n} - \underline{n}$ above threshold is usually comparable with $\underline{n}$. But for $n = \underline{n}$ the amplification would compensate the losses $\varkappa_1$¶ ; therefore, for $n = \bar{n}$ the gain will exceed the absorption coefficient $\varkappa_1$ by an amount of the order of $\varkappa_1$. Therefore, during the first stage of generation, when $n > \underline{n}$, the light energy in the

† In this chapter we present the results of [71, 72], which were obtained by the author in collaboration with V. S. Chernov.

‡ The resonator Q is defined as $\omega/v\varkappa_1$; we shall use this term as an abridged designation of the quantity which is the reciprocal of the light-energy losses $\varkappa_1$.

¶ For simplicity we shall examine a four-level diagram.

cavity increases according to the law

$$\mathcal{E} \sim \mathcal{E}_0 \exp(v\varkappa_1 t). \tag{40.1}$$

The stimulated-emission energy increases until the intense deexcitation of the excited atoms causes their number n to become less than the threshold value $\underline{n}$. For $n < \underline{n}$ the effective absorption coefficient becomes positive, and $\mathcal{E}$ begins to decrease rapidly; as a result, a sharp intensity peak is formed. From Eq. (40.1) it is evident that the light energy changes substantially during the time $1/v\varkappa_1$; thus, we obtain the following estimate for the width of the intensity peak:

$$\delta t \sim \frac{1}{v\varkappa_1}, \tag{40.2}$$

which is usually about 10^{-8} sec (i.e., it is two orders of magnitude smaller than the width of the peak for relaxation oscillations (Chap. VII).

The single pulse is characterized not only by the width of the peak but also by its position relative to the instant corresponding to the beginning of generation. Assume that generation begins at time $t = 0$. Then the constant $\mathcal{E}_0$ in Eq. (40.1) is the light energy at the initial instant of generation (i.e., the spontaneous-emission energy which falls within the solid angle $\delta\Omega$; therefore, $\mathcal{E}_0$ is proportional to the small solid angle $\delta\Omega$ within which the stimulated emission is concentrated. A time interval $t^* \sim (v\varkappa_1)^{-1}\ln(\mathcal{E}_{max}/\mathcal{E}_0)$ is required for a single pulse to develop from this priming energy according to the law (40.1). The most significant of the factors appearing in the logarithm is the quantity $1/\delta\Omega$. Thus, the intensity peak is not formed immediately after the optical shutter is opened and generation begins, but only when the time interval

$$t^* \sim \frac{1}{v\varkappa_1}\ln\frac{1}{\delta\Omega} \sim \delta t\cdot\ln\frac{1}{\delta\Omega} \tag{40.3}$$

has elapsed after the instant corresponding to the start of generation. This time interval considerably exceeds the width of the peak (40.2).

The factor indicated has great significance for the technology of obtaining a single pulse, since the optical shutter must open for a relatively long time interval δt, which exceeds $\delta t|\ln\delta\Omega|$ by approximately one order of magnitude, rather than for the short time δt (which would be very difficult to achieve). Actually, if the optical shutter can open during the time t^* required for the formation of the intensity peak, then the peak will develop for an open shutter in just the same way as it would if the shutter were to be opened instantaneously.

However, if the time t_0 during which the optical shutter opens exceeds t^*, then the peak will be formed at an instant at which the shutter has not yet had time to open completely, and only a portion of the excess reserve of excited atoms will be deexcited in the peak. For a sufficiently large t_0, as shown in [53], generation is achieved in the form of a train of peaks having a relatively large width and a weak intensity, rather than in the form of a single pulse.

Although the time allowed for the opening of the optical shutter exceeds the peak duration considerably. It is nevertheless a small quantity of the order of 0.1 μsec. A Kerr cell and the magneto-optical effect are used to produce a high-speed optical shutter. However, such shutters do not transmit sufficiently large light fluxes in the open state; therefore, for fast Q modulation extensive use is made of a rotating end mirror. If the end reflector rotates rapidly about an axis perpendicular to the optic axis of the laser, then generation is possible only in a very short time interval during which the angle between the optic axes of the reflectors does not exceed a small value ϑ of the order of a minute. Thus, the rotating reflector is equivalent to an optical shutter which opens during the time $t_0 \sim \vartheta/2\pi f$, where f is the number of revolutions per second. If, for example, f = 500 rps and $\vartheta = 1'$, then $t_0 = 10^{-7}$ sec; this value is small enough to obtain a single pulse.

The enumerated versions of optical shutters are called active shutters, in contrast with passive shutters which are opened by the generated radiation itself (Chap. XI) and cannot be opened precisely at a predetermined instant.

Many papers (for example, [9, 21, 27, 53, 56, 138, 139, 140, 141]) have been devoted to the investigation of single pulses obtained by means of active Q modulation. In theoretical papers it is usually assumed that the optical shutter opens instantaneously. [53] and [21] are exceptions; in the first of these the equation for the intensity was solved numerically, while in the second an analytical expression was derived for the leading edge of the peak (in general, the peak is not symmetrical).

In the presentation given below the results of the papers enumerated are generalized for the case of an optical shutter which opens at a finite rate according to an arbitrary law [72]. The analysis can be carried out for a four-level diagram, but the results are also applicable to a three-level laser with the proviso that in the case of a three-level diagram the quantity n and its threshold values $\bar{n}$, $n^*(t)$, $\underline{n}$ must be measured from the value $n_0/2$, where n_0 is the total volume concentration of impurity luminescence centers (for example, instead of n^* we must write $n^* - n_0/2$ in all the equations).

§ 41. Equations for Light Energy in a Resonator with Modulated Q

In a resonator with a rapidly switched Q the gain is considerably greater than it is in a conventional resonator with a fixed Q. Therefore, although for single-pulse operation modes are generated under different conditions just as in conventional operation, the difference between these conditions can nevertheless be neglected in comparison with the large gain which is identical for all modes.† Therefore, all of the modes behave practically identically, and the radiation can be treated as a single entity. This allows us to make use of the kinetic equation (25.14), as well as Eq. (25.15).

In order to consider fast Q modulation we introduce certain modifications into these equations. In Eq. (25.15) we drop the first two terms in the right side which describe the pumping and the spontaneous emission, since they do not have time to take effect during the short duration of the generation. Taking account of the dependence of the light losses on time, we write Eq. (25.13) somewhat differently: $\varkappa(\omega_0) = \varkappa_1 \varphi(t) - (\underline{n}/\underline{n})$; here $\varkappa_1\varphi(t)$ is the magnitude of the light losses at time t, where $\varphi = 1$ for an open optical shutter, and $\varphi > 1$ for a closed shutter. Using the fact that the threshold value of n is proportional to the magnitude of the losses for a four-level diagram, we write φ in the convenient form

$$\varphi(t) = \frac{n^*(t)}{n}, \tag{41.1}$$

where n^* is the threshold number of excited atoms; $\underline{n}$ is the minimum value of n^* which is achieved with the shutter open.

† According to Eq. (32.1) a short duration of the generation corresponds to large spectral width of the radiation which accommodates a large number of longitudinal modes in accordance with the discussion in §15; in this case the longitudinal intensity distribution is practically uniform.

Let us combine Eqs. (25.13)-(25.15) as we did in § 25. Solving Eq. (25.14) for $\rho(\omega, t)$ and substituting the result into (25.15), we obtain the kinetic equation for a resonator with a modulated Q:

$$y''(t) - v\varkappa_1 \frac{d\varphi}{dt} - v\varkappa_1 [v\varkappa_1\varphi(t) - y'(t)]\, e^{-\Lambda - y} = 0. \quad (41.2)$$

Here we have introduced the substitutions

$$y = v\int_0^t \varkappa(\omega_0, t)\, dt, \quad (41.3)$$

$$e^{-\Lambda} = \frac{\delta\Omega}{2\pi T\sqrt{v\varkappa_1 t^*}} \sqrt{\frac{\bar{n}}{\underline{n}}} \int_0^\infty dt \exp\left\{\int_0^t \left[\varphi(t') - \frac{\bar{n}}{\underline{n}}\right] v\varkappa_1 dt'\right\}. \quad (41.4)$$

The quantity y expresses the volume energy density directly:

$$J = \hbar\omega_0 \underline{n} \exp(-\Lambda - y). \quad (41.5)$$

The quantity (41.4), which contains the small factor $\delta\Omega$, is the small parameter of the theory (usually, $\exp(-\Lambda) \sim 10^{-5}$). As has already been said, due to the smallness of $\delta\Omega$ the ratio between the width of the peaks and the total duration of the generation is a small quantity of the order of $1/|\ln \delta\Omega| \sim 1/\Lambda$. The quantity $1/\Lambda$, along with $e^{-\Lambda}$, shall be treated as the small parameter of the theory.

Making use of this parameter, we assume that during the peak we have $\varphi(t) = \text{const} = \varphi(t^*)$, where t^* is the instant corresponding to maximum intensity. We introduce the following substitutions:

$$\bar{p} = \frac{\bar{n}}{\underline{n}}, \quad (41.6)$$

$$p = \frac{\bar{p}}{\varphi(t^*)} = \frac{\bar{n}}{\underline{n}\varphi(t^*)} = \frac{\bar{n}}{n^*(t^*)}. \quad (41.7)$$

Here p is the ratio between the initial number of excited atoms and the threshold number at the instant corresponding to the maximum intensity; $\bar{p}$ is the maximum possible value of p, which is reached if the peak is formed with the shutter completely open. Substituting $\varphi(t) = \varphi(t^*) = \bar{p}/p$ into Eq. (41.2) and converting to the convenient dimensionless time

$$\tau = \frac{\bar{p}}{p} v\varkappa_1 t, \quad (41.8)$$

we reduce the kinetic equation to the final form

$$y''(\tau) - \frac{p}{\bar{p}} [1 - y'(\tau)]\, e^{-\Lambda - y} = 0. \quad (41.9)$$

Equation (41.9), unlike the original equation (41.2), does not describe the entire generation process but only the intensity peak for a stipulated value of the parameter p. This parameter is determined by the Q-switching law which is manifested at the initial stage of generation that precedes the peak. The parameter p can be found from Eq. (41.2). Thus, the problem can be split into two parts. First, we investigate the shape of the peak for a stipulated value of the parameter p (§ 42) using Eq. (41.9), and then we find the dependence of the parameter p on the Q-switching law (§ 43) by examining Eq. (41.2).

§ 42. Shape of the Intensity Peak

It is not difficult to find the first integral of Eq. (41.9):

$$p-1+y'(\tau)+\ln\frac{1-y'(\tau)}{p}=\frac{\bar{p}}{p}e^{-\Lambda-y(\tau)}. \tag{42.1}$$

Here the integration constant was chosen while taking account of the fact that immediately before the beginning of the peak we have $n=\bar{n}$, $\varphi(t)=\varphi(t^*)=\bar{p}/p$, while the right side of Eq. (42.1) describing the stimulated emission is zero with sufficient accuracy.

Assuming that $y'=0$ in Eq. (42.1), we find the maximum value of the volume energy density†

$$J_{\max}=\hbar\omega_0\bar{n}\left(1-\frac{\ln p+1}{p}\right); \tag{42.2}$$

$J_{\max}$ increases as p increases and reaches saturation for $p \gg 1$.

Integrating Eq. (42.1), we obtain the time dependence of the intensity in parametric form:

$$\tau-\tau^*=\int_z^{1/p}\frac{dx}{x[p(1-x)+\ln x]}, \tag{42.3}$$

$$J=\hbar\omega_0\bar{n}\left(1-z+\frac{\ln z}{p}\right) \tag{42.4}$$

(z is the parameter; τ^* is the instant corresponding to the maximum intensity). This dependence is simplified in the limiting cases of small and large initial excesses above threshold.

† This result was obtained by Prokhorov [56] in somewhat different form.

For $p - 1 \ll 1$, we have

$$J = \frac{1}{2}\hbar\omega_0\bar{n}(p-1)^2\frac{1}{\left[\cosh\frac{p-1}{2}\tau\right]^2} \tag{42.5}$$

The peak has a symmetrical form which is almost Gaussian; the half-width of the peak is

$$\delta t = \frac{3.52}{(p-1)\,\bar{p}v\varkappa_1}. \tag{42.6}$$

With increasing $\bar{p}$ and p the height of the peak increases, while the half-width decreases. The shape of the single pulse differs from the shape of a free-oscillation peak (§ 27) solely in its time scale; this is connected with the fact that for $p - 1 \ll 1$ the generation of a single pulse, just as the generation of free oscillations, is accompanied by a small deviation of n from its threshold value.

Let us examine the second limiting case in which $p \gg 1$, i.e., the initial value $n = \bar{n}$ considerably exceeds the threshold value $n(t^*)$. In this case the half-width of the peak is

$$\delta t = \frac{1}{\bar{p}v\varkappa_1}(\ln p + p\ln 2). \tag{42.7}$$

The peak is sharply asymmetrical: its rise time (the first term in Eq. (42.7)) is considerably shorter than the decay time (the second term). The short duration of the intensity rise is critically related to the fact that a large initial excess above threshold corresponds to a large gain. The relatively long decay time is explained by the fact that after de-excitation of the excess reserve of excited atoms the light energy in the resonator decreases at the usual rate corresponding to the magnitude of the light losses $\varkappa_1\varphi(t^*) = \varkappa_1\bar{p}/p$ at time t^*.

Let us fix the parameter $\bar{p} = \bar{n}/\underline{n}$ which stipulates the initial value of n; we shall increase the Q-switching time within limits which provide for satisfaction of the condition $p \gg 1$. As the Q-switching time is increased, $\varphi(t^*)$ will increase, p will decrease, the peak width will decrease, and J_{max} will remain practically unchanged and equal to $\hbar\omega_0\bar{n}$ in accordance with Eq. (42.2). However, the total energy generated in the peak will decrease due to the decrease in the width of the peak.

However, if for $\bar{p} \gg 1$ the optical shutter opens fast enough so that $p = \bar{p}$, it follows that with an increase in the parameter p =

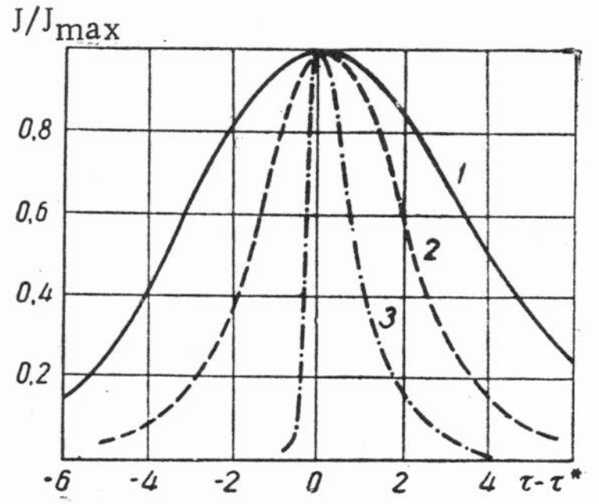

Fig. 40. Shape of the single pulse for various values of p: 1) p = 1.5; 2) p = 3; 3) p = 10.

$\bar{p}$ the quantity J_{max} increases in proportion to p, while δt decreases negligibly. For comparison purposes let us recall that for a small initial excess above threshold ($p - 1 \ll 1$) the value of J_{max} increases, while δt decreases as each of the variables p and $\bar{p}$ increases separately. The shape of the peak for various values of the parameter p is shown in Fig. 40 (the time expressed in $p/\bar{p}v\varkappa_1$ units is plotted along the axis of abscissas). See ref. [142] for a discussion of pulse shape.

§43. Effect of a Finite Q-switching Rate and Rate and Pumping Nonuniformity on the Shape of the Peaks

The Effect of the Q Switching Law. Above we showed that the shape of the intensity peak and its maximum value are completely determined by the parameters $\bar{p}$ and p, of which the second is not known in advance and is determined by the Q-switching law. Let us go over to the solution of the second part of the problem – finding the parameter p.

In order to find p it is necessary to know the instant t* corresponding to the maximum intensity, or the practically equivalent instant t_i which corresponds to the beginning of the peak and is defined as follows:

$$J(t_i) = \varepsilon J_{max} \tag{43.1}$$

Here $\varepsilon \ll 1$ is a small parameter which is chosen in such a way that

$$|\ln \varepsilon| \ll \Lambda \tag{43.2}$$

(we recall that $\Lambda \gg 1$). Because of the inequality (43.2), ε drops out of the final result.

In order to find the instant t_i we shall consider the kinetic equation separately in the time domain preceding the peak ($t \leq t_i$) and in the time domain of the peak ($t \geq t_i$). In the domain preceding the peak it is possible to drop the last term in Eq. (41.2), which is proportional to the stimulated emission; the intensity of this emission is a negligibly small quantity outside the domain of the peak. Then the equation can easily be integrated, and its solution has the form

$$J = \hbar\omega_0 \underline{n} \exp\left\{-\Lambda + v\varkappa_1 \int_0^t [\bar{p} - \varphi(t')]\, dt'\right\}. \tag{43.3}$$

Let us substitute this solution into the left side of Eq. (43.1), while (42.2) for J_{max}, which is determined from an analysis of the kinetic equation in the domain of the peak, is substituted into the right side. As a result, we obtain the equation for $t_1 \cong t^*$ (i.e., actually for p):

$$\frac{1 + \ln p}{p} = 1 - \frac{1}{\bar{p}} \exp\left\{-\Lambda - v\varkappa_1 \int_0^{t^*(p)} [\varphi(t) - \bar{p}]\, dt\right\}. \tag{43.4}$$

Here $t^*(p)$ is a function which is stipulated by the relationship $p\varphi(t^*) = \bar{p}$. The parameter ε has dropped out of this equation due to the inequality (43.2). For simplicity we have also replaced t_i by the closely similar value t^* using the smallness of the width of the peak in comparison with the total duration of the generation. In both cases we neglect the small parameter Λ^{-1} in comparison with unity.

Let us examine certain particular cases.

1. The optical shutter opens according to the exponential law

$$\varphi(t) = 1 + (\bar{p} - 1) \exp\left(-\frac{t}{t_0}\right), \quad 1 < \varphi \leqslant \bar{p} = \varphi(0)$$

(t_0 is the Q-switching time, and the generation begins at time $t = 0$). Expressing t^* in terms of p on the basis of this expression and substituting into Eq. (43.4), we obtain the equation for p:

$$\frac{1 + \ln p}{p} = 1 - \frac{1}{\bar{p}} \exp\left\{-\Lambda + v\varkappa_1 t_0 \left[(\bar{p} - 1)\, \mathrm{n} \frac{p(\bar{p} - 1)}{\bar{p} - p} - \frac{\bar{p}(p - 1)}{p}\right]\right\}.$$

This equation is easily solved if the optical shutter is almost completely open at the instant at which the peak is formed. Then $\bar{p}$ –

$p \ll \bar{p}$; making use of this, we find

$$p = \bar{p}\left\{1 - \frac{\bar{p}-1}{e}\exp\left[-\frac{\Lambda + \ln(\bar{p}-1-\ln\bar{p})}{v\varkappa_1 t_0(\bar{p}-1)}\right]\right\}.$$

From this it is evident that for the condition

$$t_0 \ll \frac{\Lambda}{v\varkappa_1(\bar{p}-1)}$$

the value of p coincides with its maximum possible value $\bar{p}$, i.e., the shutter opens practically instantaneously. A finite Q-switching rate has an effect only at large values of t_0 which are comparable with the quantity $\Lambda/v\varkappa_1(\bar{p}-1)$ (i.e., with the time interval required to form the peak; see § 40).

2. The optical shutter opens according to the power law

$$\varphi(t) = 1 + (\bar{p}-1)\left(1 + \frac{t}{t_0}\right)^{-q}, \quad q > 0, \quad q \sim 1.$$

For $\bar{p} - p \ll \bar{p}$ similar calculations lead to the following result:

$$p = \bar{p}\left\{1 - (\bar{p}-1)\left[\frac{v\varkappa_1 t_0(\bar{p}-1)}{\Lambda + \ln(\bar{p}-1-\ln\bar{p})}\right]^q\right\}.$$

All of the qualitative conclusions drawn in the previous case remain in force.

3. In the case of a rotating reflector the Q is switched during a short time interval and reaches a maximum at a certain time t_0 near which it varies according to a quadratic law:†

$$\varphi(t) = 1 + (\bar{p}-1)\left(\frac{t}{t_0} - 1\right)^2, \quad \varphi(0) = \bar{p}, \quad \varphi(t_0) = 1. \tag{43.5}$$

Solving Eq. (43.4) for $\bar{p} - p \ll \bar{p}$, we find

$$p = \bar{p}\left[1 - \frac{4}{9}(\bar{p}-1)\left(1 - \frac{t_c}{t_0}\right)^2\right], \tag{43.6}$$

where

$$t_c = \frac{3}{2(\bar{p}-1)v\varkappa_1}[\Lambda + \ln(\bar{p}-1-\ln\bar{p})]. \tag{43.7}$$

† The time origin is chosen to be the instant at which generation begins, since the initial value of n, which is equal to $\bar{n}$, coincides with the threshold value $n^*(0) = \underline{n}\,\varphi(0)$. From this we find that $\varphi(0) = \bar{p}$.

The instant corresponding to the maximum intensity is equal to

$$t^* = t_0\left[1 + \sqrt{\frac{\bar{p} - p}{\bar{p}(\bar{p} - 1)}}\,\mathrm{sgn}(t_c - t_0)\right]. \tag{43.8}$$

From the equations given above it is evident that if the Q-switching time t_0 coincides with t_c, then the instant corresponding to maximum intensity coincides with the instant corresponding to maximum Q, while p reaches its upper boundary $\bar{p}$. For $t_0 > t_c$ the intensity peak precedes the instant corresponding to maximum Q, while for $t_0 < t_c$ it occurs after this instant; in both case $p < \bar{p}$.

Experimental data on single pulses are available only for the case of a plane-parallel resonator [9, 53]. These data provide qualitative confirmation of the theory, but the theoretically predicted length of the single pulse is often noticeably shorter than the experimental value. This is explained by the nonuniform initial distribution of the concentration of excited atoms over the cross section of the active element, which is connected with the nonuniform pumping. The spatial nonuniformity of the gain leads to a situation in which the intensity increases at different rates at different points on the cross section, and the instants corresponding to the maximum intensity do not coincide for the different points.

The Effect of Spatial Nonuniformity. The dynamics of the development of a single pulse in a plane-parallel resonator was considered by Letokhov and Suchkov [41] who took account of the nonuniform distribution of the population inversion over the cross section. This paper was based on the wave equation for the electromagnetic field in a resonator containing an active medium; in the notation which we have adopted this equation has the form (25.1). In § 25 we assumed that the electromagnetic field in the resonator behaves as a single entity (i.e., the amplitude A of the field and the effective absorption coefficient $\varkappa$ do not depend on the coordinate); on this assumption we obtain the balance equation from the wave equation. In [41], which takes account of the spatial nonuniformity of the gain, the electromagnetic field was written in the form $A(x, t) \exp(i\omega_0 t)$, where A is the amplitude and depends on both time and the transverse coordinate a. When this expression is substituted into the wave equation for the field, as in § 25, it is necessary to take account of the fact that the characteristic time for the variation of the amplitude A exceeds $1/\omega_0$ considerably, and this allows the second derivative of the amplitude with respect to time to be dropped. As a result, the wave equation takes the form

$$\dot{A} = \frac{iv^2}{2\omega_0} A''_{xx} - \frac{i\varkappa\omega_0}{2k} A. \tag{43.9}$$

Equation (43.9) must be supplemented by the equation for $\varkappa$, which we shall not write out because it is similar to the equations of § 25.

In [41] Eq. (43.9) was solved on a computer by expanding the field in natural oscillations of the resonator. Figure 41 shows the space-time picture of the development of a single pulse [41], as calculated for characteristic values of a ruby laser (l = 50 cm, a diameter of the active element $2a$ = 7mm, $\lambda = 7 \cdot 10^{-5}$ cm, light losses of approximately 60% for a single pass, and an initial population inversion which decreases smoothly by a factor of two from the center of the cross section to the edges). The curves 2, 3, 4 show the development of the single pulse at different points on the cross section: at distances of $a/2$ and $3a/4$ from the center and at the center of the cross section, respectively; curve 1 characterizes the time dependence of the intensity of the resultant pulse. The quantity x_0 designates the half-width of the generation region (i.e., the radius of the luminescence region in which the intensity of the generated light exceeds one half of the maximum value).

From Fig. 41 it is evident that the length of the resultant pulse noticeably exceeds the length of the pulse which develops at a fixed point of the cross section. Thus, the nonuniformity of the population inversion, which is connected with the nonuniform distribution of the pump intensity, leads to a noticeable increase of the duration of the single pulse in the case of a plane-parallel resonator.

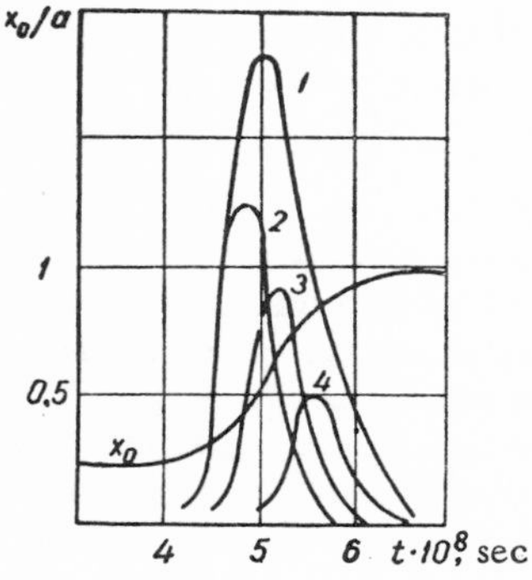

Fig. 41. Spatial structure of the single pulse.

In [72] an analytical estimate is derived of the single-pulse time broadening which is connected with the nonuniform distribution of the population inversion in a plane-parallel resonator. This estimate was obtained by solving the equation in the time domain preceding the peak, where the population inversion retains its initial value and Eq. (43.9) is linear. This equation can be solved in the Laplace representation by means of the WKB method. As a result, the following estimate is obtained for the single-pulse time broadening connected with nonuniformity of the gain:

$$\delta t_{\text{H}} \approx \min\begin{cases} \dfrac{|\varkappa(0)-\varkappa(a)|}{|\varkappa(0)|}t^{*}, \\ \dfrac{1}{v}\sqrt{\dfrac{2k}{|\varkappa(0)|}}\displaystyle\int_0^a \sqrt{1-\dfrac{|\varkappa(r)|}{|\varkappa(0)|}}\,dr. \end{cases} \tag{43.10}$$

Here r is the distance of the point from the axis of the cylindrical rod; $\varkappa(0)$ is the effective absorption coefficient at the center of the cross section; $\varkappa(a)$ is the effective absorption coefficient near its boundary (it is assumed that $\varkappa(0) < \varkappa(a) < 0$). The time broadening of the single pulse decreases as the wave vector and the cross-section diameter decrease (i.e., as the role played by the diffraction phenomena which facilitate mixing of the radiation increases).

The estimate (43.10) is applicable for the condition that δt_H exceeds the length of the single pulse in a resonator having a uniform distribution of the population inversion. For the parameter value values used in [41] Eq. (43.4) yields $\delta t_H = 11$ nsec and is in sufficiently good agreement with Fig. 41.

What has been said above applies to a plane-parallel resonator. However, in the case of a spherical resonator (or a resonator containing lenses) the gain nonuniformity has no effect on the duration of the single pulse, since the nonuniform distribution of the population inversion is averaged over the cross section due to the transverse motion of the light (Chap. VI).† Therefore, in the case of a

†This is valid for the condition that the amplification is a maximum near the boundaries of the generating region of the cross section; in this case a practically uniform development of the single pulse is provided for by the advance development of high transverse modes whose field encompasses practically the entire generating region. However, if the amplification is a maximum near the center of the cross section, then the low transverse modes having fields which are localized in the central region of the cross section develop in advance; this can lead to a substantial increase in the length of the single pulse [108].

spherical resonator or a resonator containing lenses the theory which has been developed in this chapter and is based on the assumption of spatial uniformity of the system retains its validity regardless of the initial distribution of the population inversion. However, in the case of a plane-parallel resonator this theory is quantitatively applicable only for the condition of a sufficiently uniform distribution of the pump intensity over the cross section.

The length of the single pulse obtained in experiments with a rapidly switched Q is measured in tens of nanoseconds. For certain applications shorter light pulses are required. An additional reduction in the length of the light pulse can be achieved by passing it through a nonlinearly amplifying medium. A theoretical and experimental investigation of this problem was carried out in [7, 8, 10].

The essence of the effects which are connected with the transmission of a light pulse through a nonlinearly amplifying medium resides in the following. The initial pulse appearing at the input of the amplifier has a relatively long leading edge; strictly speaking, the duration of the leading edge is equal to the interval between the instants t^* corresponding to the maximum intensity and the instant at which generation begins (the latter instant coincides with the Q-switching instant). The magnitude of this interval exceeds the half-width of a single pulse considerably (§ 40). The light emitted at the early instants in time, which precede the peaks, depletes the population inversion stored in the medium while being amplified in it; therefore the light emitted by the laser at subsequent instants in time (and, in particular, at the instant t^*) travels through a medium having a gain which is considerably lower than the initial value. This leads to a shifting of the intensity maximum toward earlier times, right up to the instant corresponding to the beginning of generation; in other words, the maximum is propagated at a velocity exceeding the velocity of light. Simultaneously, the pulse narrows, provided only that its initial leading edge is sufficiently steep. Since the effects indicated are essentially connected with the nonlinear nature of the amplification, it follows that their observation requires a sufficiently large power in the initial pulse.† In [7] the contraction of the pulse length from 8.7 to 4.7 nsec was observed experimentally for its passage through a three-stage amplifier.

†See Chap. XII for a discussion of pulse velocity > c in low-power lasers.

§ 44. Forced Oscillations of the Intensity

Let us examine forced oscillations of the stimulated-emission intensity which are caused by weak periodic Q modulation. Let us write the light-energy losses referred to a unit length in the form

$$\varkappa_1\left[1+\sigma\sin\frac{2\pi t}{t_0}\right], \tag{44.1}$$

where σ and t_0 are the depth and period of the modulation. The modulation depth σ shall be assumed small.

We write the kinetic equation (25.9) while taking account of the Q-modulation law (44.1); for simplicity we neglect the spontaneous emission retained in the resonator in this equation and place† $\varkappa(\omega) = \varkappa(\omega_0)$. Using $J = \int \rho\,(\omega,\, t)d\omega$ to designate the volume energy density, we have

$$\dot{J} = -v\dot{\varkappa}(\omega_0)J = v\varkappa_1 J\frac{n}{n^*} - v\varkappa_1\left(1+\sigma\sin\frac{2\pi t}{t_0}\right)J. \tag{44.2}$$

Combining this equation with the kinetic equation (25.15), we obtain the equation for the stimulated-emission energy:

$$\ddot{w} + \frac{\dot{w}}{T}(1+\zeta) + \frac{v\varkappa_1\zeta}{T}w = \frac{2\pi\sigma v\varkappa_1}{t_0}e^{\frac{2\pi i t}{t_0}}. \tag{44.3}$$

Here w designates the magnitude of the deviation of the light energy from the value $(N - N^*)/v\varkappa_1$ corresponding to continuous operation with a fixed Q, i.e.,

$$J = \frac{N-N^*}{v\varkappa_1}(1+w), \quad |w| \ll 1. \tag{44.4}$$

Equation (44.3) is linearized with respect to the small quantity w and with respect to the small parameters σ and $1/v\varkappa_1 T$; here the modulation period t_0 is assumed to be comparable with the period of small free oscillations. Thus, Eq. (44.3) describes only small forced oscillations of the stimulated emission.

Equation (44.3) is in no way different from the equation for forced oscillations of a mechanical system under the application of

†This is equivalent to neglecting the ω-damping mechanism of the free oscillations (§28), which does not play an essential part for a solid angle $\delta\Omega$ which is not too large.

an external periodic force in the presence of friction. It is not difficult to write the solution of this equation which describes forced oscillations:

$$w = \sigma \frac{t_0}{\Delta t_0} \sqrt{\frac{v\varkappa_1 T}{\zeta}} \frac{\exp\left(\frac{2\pi i t}{t_0}\right)'}{\left(\frac{t_0}{\Delta t_0}\right)^2 - 1 + \frac{i t_0}{\Delta t_0} \cdot \frac{1+\zeta}{\sqrt{v\varkappa_1 \zeta T}}}. \tag{44.5}$$

Here

$$\Delta t_0 = 2\pi \sqrt{\frac{T}{v\varkappa_1 \zeta}} \tag{44.6}$$

is the period of small free intensity oscillations. When this period coincides with the modulation period, resonance occurs, and the amplitude of the forced intensity oscillations reaches a maximum.

The solution of (44.5) describes only small forced oscillations. However, along with small oscillations it is also possible to have strong forced oscillations with a large amplitude and sharp intensity peaks, which are analogous to the strong free oscillations considered in §27. Such forced oscillations are possible if the modulation period t_0 exceeds the period of small free oscillations Δt_0; this provides for resonance with strong free oscillations whose period is longer than Δt_0 (§ 27). Depending on the modulation depth σ, either small oscillations described by the solution (44.5) or forced oscillations having a large amplitude are realized for $t_0 > \Delta t_0$ †.

An investigation of forced oscillations with a large amplitude can be carried out using the smallness of the ratio between the width of the intensity peaks and the period [71]. Omitting the calculations, which in general are analogous to those carried out in §§ 42 and 43, we present the final result.

The character of the forced oscillations is determined by the modulation depth σ. Let us examine three ranges of the values of σ.

1. A small modulation depth:

$$\sigma < \frac{8\pi^2}{3} \left(\frac{t_0}{\Delta t_0}\right)^2 \frac{\sqrt{\zeta(\zeta+1)}}{v\varkappa_1 T} \tag{44.7}$$

† Such an ambiguity of the solution is a general property of the equation for nonlinear oscillations in the presence of a periodic constraining force [37].

(the right side of this inequality is approximately equal to 10^{-4}). Regardless of the modulation period, small oscillations which can be described by the solution (44.5) occur.

2. The intermediate range:

$$\frac{8\pi^2}{3}\left(\frac{t_0}{\Delta t_0}\right)^2 \frac{\sqrt{\zeta(\zeta+1)}}{v\varkappa_1 T} < \sigma \ll \sqrt{\frac{\zeta}{v\varkappa_1 T}}\,. \tag{44.8}$$

For $t_0 < \Delta t_0$ small oscillations (44.5) occur; the smallness of the oscillations derives from the inequality (44.8) and Eq. (44.5). However, for $t_0 = \Delta t_0$, for which resonance exists, the amplitude of the oscillations described by Eq. (44.5) is not small (so that this equation is, strictly speaking, inapplicable). The resonance condition is also not violated for $t_0 > \Delta t_0$, since the modulation period t_0 coincides with the period of the free oscillations for which the amplitude $Y \gg 1$ and the period $\Delta t > \Delta t_0$ (§ 27). Because of this resonance, the forced oscillations have a large amplitude for $t_0 > \Delta t_0$. It can be shown [71] that the forced oscillations undergone by the stimulated emission for $t_0 > \Delta t_0$ are in no way different from the free oscillations (considered in § 27) having an amplitude Y which is determined from the condition for coincidence of the modulation period t_0 with the period of the free oscillations. The dependence of the period of the free oscillations on amplitude is expressed asymptotically by Eq. (27.17), which can be written as $\Delta t = \Delta t_0\sqrt{8Y}/2\pi$. Setting the period Δt of the free oscillations equal to the modulation period t_0, we obtain the asymptotic dependence of the amplitude of the forced oscillations on the modulation period:

$$Y = \frac{\pi^2}{2}\left(\frac{t_0}{\Delta t_0}\right)^2. \tag{44.9}$$

This equation is applicable only for sufficiently large values of t_0. The dependence of Y on the modulation period t_0 for any values of t_0 can be plotted by means of Fig. 30; the graph of this dependence is shown in Fig. 42.

In the shape and height of the peaks and the length of the period the forced intensity oscillations do not differ from the free oscillations (§ 27), whose amplitude Y is stipulated by Eq. (44.9) or Fig. 42. Thus, the laser radiation in essence undergoes free oscillations, and the role played by the modulation reduces solely to fixing their period (i.e., actually their amplitude).

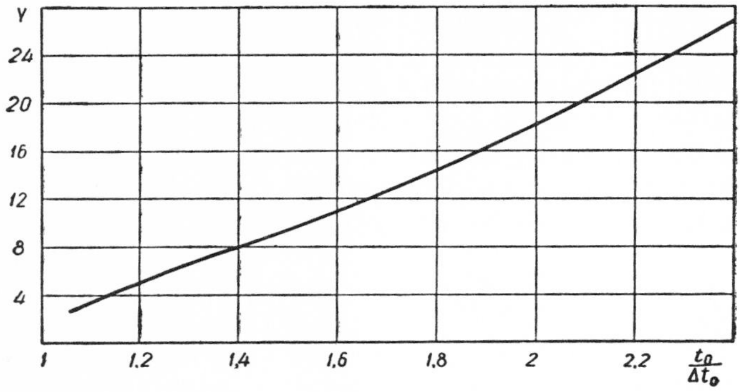

Fig. 42. Dependence of the amplitude of the forced oscillations on the modulation period $t_0/\Delta t_0$.

3. A relatively large modulation depth:

$$\sigma \gtrapprox \sqrt{\frac{\zeta}{v\varkappa_1 T}}\,. \tag{44.10}$$

What was said in subsection 2 remains in force in the range $t_0 > \Delta t_0$. However, in the range $t_0 < \Delta t_0$, as is evident from Eq. (44.5), the oscillations are not small; thus, this equation is, strictly speaking, inapplicable. Without going into a detailed analysis of forced oscillations for $t_0 < \Delta t_0$, we confine ourselves to the comment that these oscillations are dissimilar from free oscillations having a large amplitude.

Thus, in the range $t_0 > \Delta t_0$ the shape of the forced oscillations is determined by the modulation period rather than by the modulation depth for a sufficiently large modulation depth σ. Physically it is connected with the fact that the Q is practically constant over a narrow intensity peak; therefore, the shape and height of the peaks are determined solely by the number of atoms which are deexcited within the peaks (this number is equal to the number of atoms excited by the pumping during a period and is not dependent on the modulation depth). Only the choice between the two types of oscillations which are possible for $t_0 > \Delta t_1$ – small forced sinusoidal oscillations or strong oscillations with narrow intensity peaks – is dependent on the modulation depth.

We note the high sensitivity of the generated radiation to modulation even when the modulation has a shallow depth. For example, for a ruby or neodymium — glass laser the parameter $\zeta / h \varkappa_1 T$ appearing in the criterion for the appearance of strong forced oscillations is equal to approximately 10^{-5}. Such a weak Q modulation affects the generation kinetics for the reason that the insignificant change in the effective absorption coefficient connected with it leads to a noticeable change of the intensity during a relatively long period of oscillation.

The analysis carried out in this section is based on the kinetic equation, which is justified only if the stimulated emission in the resonator behaves as a single entity. The criterion for the applicability of the kinetic equation and its solution describing free oscillations was considered in § 29. This criterion also remains valid for that solution of the kinetic equation which describes the forced oscillations of the intensity. Therefore, the results obtained in the present section are applicable to those lasers whose radiation undergoes proper free oscillations in the absence of modulation.† However, it is evident that proper forced oscillations can also be observed without this restriction for a sufficiently large modulation depth (exceeding the left side of the inequality (44.8), but small in comparison with unity). In fact, Q modulation with a sufficient depth must lead to matched oscillations of the entire stimulated emission in the resonator.

In order to obtain proper forced oscillations it is necessary to place a restriction on the modulation period t_0:

$$t_0 < \Delta t_{\max} = \sqrt{\frac{8 Y_1 T}{v \varkappa_1 \zeta}}, \tag{44.11}$$

where Y_1 is the maximum amplitude of the free oscillations and is stipulated by Eq. (28.4); $\Delta t_{\max}$ is the maximum period of these oscillations (we recall that the amplitude and period of the free oscillations, unlike those of the forced oscillations, decrease with time due to damping). The restriction (44.11) is connected with the fact that for forced oscillations to exist the stimulated-emission energy which has reached the investigated intensity peak from the previous peak must not be too small; namely, this energy must exceed the spontaneous-emission energy emitted within the solid angle $\delta\Omega$ during the investigated peak itself. If this condition is

† This restriction does not extend to §41-43, which were devoted to a consideration of single pulses, since in view of the large amplification the generated radiation behaves practically as a single entity.

not satisfied, then the intensity peak will develop from the spontaneous emission rather than from the stimulated emission which performs forced oscillations; therefore, the radiation will be generated in the form of isolated peaks which are analogous to those considered in § § 41 and 43.† The stimulated-emission energy remaining in the resonator in the interval between peaks decreases exponentially with an increase of the period; it is not difficult to show that for $t_0 = \Delta t_{max}$ it becomes comparable with the spontaneous-emission energy.

Conclusions

1. A single pulse having a length of the order of $(v\varkappa_1)^{-1} \sim 10^{-8}$ sec is generated in a resonator having a rapidly switched Q. The shape of the single pulse is determined by the parameters $\overline{p}$ and p $(1 < p < \overline{p})$, where $\overline{p}$ or p is the ratio between the initial number of excited atoms and their threshold value at the instant corresponding to the maximum Q or the maximum intensity, respectively. The maximum intensity and the total energy generated in the peaks increase as $\overline{p}$ and p increase; the width of the peak decreases as $\overline{p}$ increases. If p is increased for a stipulated $\overline{p}$ (i.e., if the instants corresponding to maximum intensity and maximum Q are brought closer together), then the width of the peak will decrease for small p and increase for $p \gg 1$.

2. The peak is not formed immediately after Q switching, but after a delay $t^* \sim |\ln \delta\Omega|/v\varkappa_1$ which considerably exceeds the width of the peak. If the Q-switching time $t_0 < t^*$, then the peak is formed when the optical shutter is open, and thus $p = \overline{p}$ as in the case when the shutter opens instantaneously. However, if $t_0 > t^*$, then the shutter does not have time to open by the time the peak is formed, and $p < \overline{p}$.

3. An optimal speed exists for the rotation of the reflector used for fast Q switching.

4. Periodic Q modulation with a small depth leads to proper forced intensity oscillations under certain conditions, and in certain cases the forced oscillations differ from free oscillations having a large amplitude (Chap. VII) solely by the absence of damping and the presence of a fixed period.

† If t_0 exceeds Δt_{max} sufficiently, then several peaks having different heights can be generated during one modulation period.

Chapter XI

PASSIVE Q MODULATION

Q-modulation of an optical cavity can be accomplished by means of a passive optical shutter which is opened due to the action of the generated radiation itself. A passive optical shutter can consist of an absorbing medium placed in the laser cavity [15, 38, 49, 81, 100, 118, 143, 144, 145]. When generation begins, the absorbing centers are transferred to an excited state, and the medium becomes transparent (i.e., the shutter is opened).

Compared with a conventional (active) optical shutter, a passive optical shutter has the shortcoming that a single pulse cannot be obtained precisely at a stipulated instant in time. On the other hand, a passive shutter offers substantial advantages: extreme simplicity of operation and ease in fabrication, as well as a large variety of absorbing media.

§45. The Mechanism of Passive Q Modulation

The absorbing medium which is used for passive Q modulation must satisfy certain requirements. In order to prevent the optical shutter from being too inert, the number of absorbing centers must be much smaller than the number of excited active luminescence centers. From this it follows that the oscillator force must be much greater for the absorbing centers than for the active centers (otherwise, the absorbing centers do not produce any noticeable absorption). Then the absorption band of the passive shutter must be broad enough so as to overlap the spectral line of the laser. Finally, as will be shown below, the absorbing medium must luminesce with a sufficiently high quantum yield.

The behavior of the optical shutter depends essentially on what occurs with the absorbing centers after their excitation. A passive shutter has a low inertia (i.e., it opens rapidly and completely) if the absorbing centers have a metastable level to which they transfer after excitation. In this case it is sufficient to excite the absorbing medium once in order to open the optical shutter for the entire generation time.

However, usually the absorbing centers do not have such a metastable level, and it is necessary to excite them repeatedly as they return to the ground state in order to maintain them in the excited state. This leads to an increase in the inertia of the optical shutter. The fact that the excited absorbing centers are deactivated (i.e., transferred to the ground state) both with the emission of light and nonradiatively (see §2) substantially affects the inertia of the shutter. If T_a is the time required for a spontaneous optical transition in an absorbing center, then the total probability of deactivation of an absorbing center per unit time is

$$w_a = \frac{1}{T_a K_a}, \tag{45.1}$$

where K_a is the quantum luminescence yield of the absorbing medium (§ 2).

As w_a increases, the absorbing centers are deactivated more frequently, and they must be transferred back to the excited state more frequently. Therefore, the inertia of the optical shutter increases with a decrease of the quantum luminescence yield K_a. However, the inertia of the optical shutter does not depend on the quantity T_a (of course, within the specified limits expressed by condition (46.5)). Let us actually imagine that T_a has been reduced by a factor of two; then it is sufficient to use half the number of absorbing centers in order to obtain the previous absorption. But since the deactivation probability of the excited absorbing centers simultaneously increases by a factor of two, it follows that the number of absorbing centers which must be transferred to the excited state per unit time remains the same.

Let us consider the operating mechanism of a passive optical shutter in greater detail. At first we shall assume that the transmission of the shutter is close to unity. When the number of excited active centers reaches the threshold value $\bar{n}$ corresponding to a closed shutter generation begins in the space between end mirrors (Fig. 43a). The small light energy which develops in the resonator opens the shutter slightly; this leads to an insignificant lowering of the threshold number of excited atoms, and an excess number of them are now deexcited. Under these conditions the light

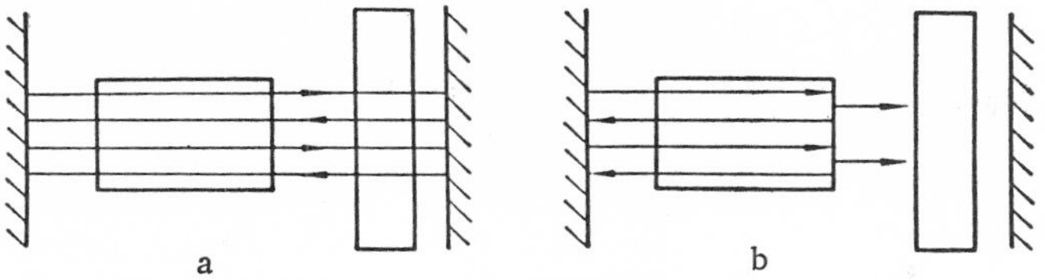

Fig. 43. Generation in the entire space between mirrors for a large transmission of the shutter (a), or generation in the space between the mirror and the endface of the rod for a low transmission (b).

energy in the resonator increases and transfers an additional number of absorbing centers to the excited state. As a result, the shutter becomes more transparent (opens further), etc. Below we shall show that either the development of this process is halted at an early stage due to the high inertia of the optical shutter (and in this case the shutter opens only to an insignificant degree), or the optical shutter opens completely and the entire excess number of excited active centers $\bar{n} - n$ ($\bar{n}$ or n represents the threshold number of excited atoms for a closed or completely open optical shutter, respectively) is deexcited.

However, if the absorbing medium is insufficiently transparent (the transmission is considerably less than unity), then for an open optical shutter generation develops in the space between the mirror and the end face of the active sample (Fig. 43b) since the absorbing medium is practically opaque to light [15]. The optical shutter is opened by the light which exits from this resonator. Note that a small initial transmission of the optical shutter leads to a high inertia of the shutter. Actually, the transmission of the shutter is equal to exp (− D), where D is the optical density and is proportional to the number of absorbing centers. From this it is evident that the derivative of the transmission with respect to the number of absorbing centers is proportional to exp (−D) and is a small quantity for large D.

Below we examine an optical shutter with a transmission comparable with unity. The calculations will be carried out for a four-level diagram, but the results are also applicable to a three-level diagram if the substitutions in (9.3) are taken into account; here in the case of a three-level diagram it is necessary to measure n from the value $n_0/2$ and to make the corresponding substitutions $\overline{n} \to \overline{n} - (n_0/2)$, $\underline{n} \to \underline{n} - (n_0/2)$ (n_0 is the total number of luminescence centers per unit volume).

§46. The Equation for a Passive Optical Shutter

Assume that active and absorbing samples having the lengths l_0 and l_a are introduced into a resonator formed by mirrors spaced a distance l apart. In order to exclude the length of the sample from consideration we shall assume that the transmission of the absorption medium is comparable with unity, so that generation develops in the space between the end mirrors (Fig. 43a). Let us imagine that both the active and absorbing media are uniformly distributed in this space, and let us use n to designate the number of excited active centers per unit volume of the uniformly distributed active medium. Similarly, we introduce a reduced volume concentration n_a of absorbing centers into the unexcited absorbing medium. Obviously, the reduced concentrations n and n_a are the true values of the concentrations multiplied by l_0/l and l_a/l, respectively.

Below we shall make use of only the reduced concentrations and reduced absorption coefficients without special comment.

In order to write the kinetic equation for a resonator containing an active medium it is first necessary to calculate the effective absorption coefficient, which appears in this equation. The effective absorption coefficient can be represented in the form

$$\varkappa(\omega_0) = \varkappa_1 - Bn + \mu_a x. \tag{46.1}$$

Here

$$\mu_a = \frac{9.4v^2 n_a \chi_a}{T_a \omega_0^2 \Delta\omega_a} \tag{46.2}$$

is the absorption coefficient of the unexcited absorbing medium at the laser frequency ω_0; x is the fraction of absorbing centers in the ground state; B is a constant:

$$B = \frac{9.4v^2}{T\omega_0^2 \Delta\omega}; \tag{46.3}$$

χ_a is the ratio between the absorption coefficient of the absorbing medium at the laser frequency ω_0 and the maximum absorption

† In the case of a three-level laser it is necessary to replace χ_a by the quantity $\chi_a/2$ in all of the equations.

coefficient;† $\Delta\omega_a$ is the half-width of the absorption band; T_a is the time required for a spontaneous optical transition in the absorbing centers (the subscript a indicates the parameters associated with the absorbing medium). For simplicity, all of the optical bands are assumed to be Gaussian.

The quantity x, which describes the effect of the light on the optical shutter, is not difficult to calculate if the following comment is taken into account. As has already been said, the number of active centers n must considerably exceed the number of absorbing centers n_a. On the other hand, the absorption $\mu_a x$ of the shutter must be comparable with the amplification Bn. From this it follows that $B \ll \mu_a/n_a$, or † taking Eqs. (46.2), (46.3) into account we have

$$T_a \ll T\chi_a\Delta\omega/\Delta\omega_a \ll T. \tag{46.4}$$

Thus, the time required for the spontaneous transition in the absorbing centers must be fairly short. In order to simplify the calculations we shall assume that the following condition is also satisfied along with the inequality (46.4):

$$T_a \ll \frac{t_0}{K_a}, \quad \text{i.e.,} \quad w_a t_0 \gg 1, \tag{46.5}$$

where $K_a < 1$ is the quantum luminescence yield of the absorbing centers; w_a is the probability of their deactivation per unit time; t_0 is the duration of the process by which the optical shutter opens. Below we shall see that for a passive shutter t_0 is much larger than for an active shutter and is about 10^{-6} to 10^{-5} sec.‡ Therefore the condition (46.5) is certainly satisfied by absorbing centers having an allowed transition.

From inequality (46.5) it follows that the quantity x, which is equal to the fraction of absorbing centers in the ground state, is determined by the value of the light energy at this same instant. In order to find x it is necessary to set the number of absorbing centers excited per unit time equal to the number of reverse transitions [i.e., it is necessary to place $J v \mu_a x/\hbar\omega_0 = n_a w_a (1 - x)$). Substituting the value of x found from this into Eq. (46.1), we find the active absorption coefficient as a function of the volume density

† The operating transitions usually correspond to luminescence with a small width $\Delta\omega$, while the absorption band of the shutter must be broad in order to overlap the laser line. From this we have $\Delta\omega \ll \Delta\omega_a$.

‡ This leads to a narrower spectral width of the radiation (see §31).

J of the light energy:

$$\varkappa(\omega_0) = \varkappa_1 - Bn + \mu_a \left[1 - \frac{1}{1 + \hbar\omega_0 n_a w_a / Jv\mu_a}\right]. \qquad (46.6)$$

Let us substitute (46.6) into the kinetic equation (25.9) after first integrating both of its sides with respect to frequency within the limits of the spectral width $\delta\omega$ of the laser; here we make use of the smallness of $\delta\omega$ and place $\varkappa(\omega) = \varkappa(\omega_0)$ in the equation.† Going over to the volume density of the light energy $J = \int \rho(\omega, t)\, d\omega$, we have

$$\dot{J} = \delta N_{sp} + Jv\left\{Bn - \varkappa_1 - \mu_a\left(1 - \frac{1}{1 + \dfrac{\hbar\omega_0 n_a w_a}{Jv\mu_a}}\right)\right\}, \qquad (46.7)$$

where $\delta N_{sp} \cong N^* \delta\Omega\delta\omega / 4\pi\Delta\omega$ is that portion of the spontaneous-emission power which is retained in the resonator. Equation (46.7) must be supplemented by the equation for the number of excited active centers (25.15), which can be rewritten thus:

$$\hbar\omega_0 \dot{n} = N - \hbar\omega_0 n / T - vBnJ. \qquad (46.8)$$

The threshold number of excited active centers for an open or closed shutter ($\underline{n}$ or $\overline{n}$, respectively) is determined from the condition for the vanishing of the effective absorption coefficient (46.1). Taking account of the fact that for an open shutter x = 0, while x = 1 for a closed shutter, we find the relationship between the threshold values of n and the constants μ_a and B:

$$\varkappa_1 = B\underline{n}, \quad \varkappa_1 + \mu_a = B\overline{n}. \qquad (46.9)$$

§ 47. A Passive Shutter Open Either Slightly or Completely

Equations (46.7), (46.8) for J and n form a closed system. However, they turn out to be rather complex; in order to simplify them it is necessary to trace the development of the generation process [61].

Generation begins at time t = 0 when the value of n reaches the threshold value $\overline{n}$ due to the action of the pumping (Fig. 34).

† For operation with Q-modulation this is allowable because of the large value of $[\varkappa(\omega_0)]$.

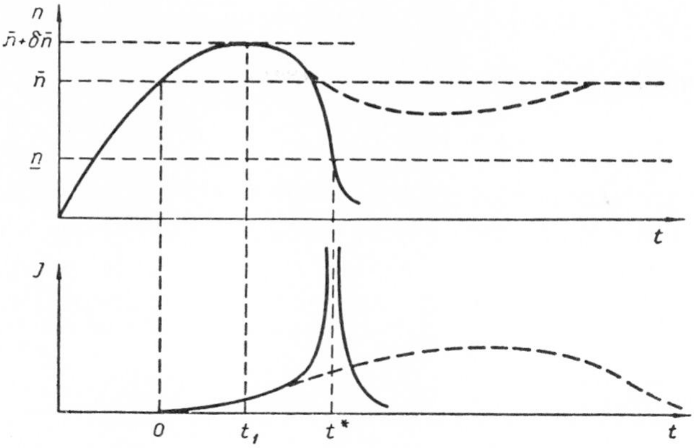

Fig. 44. Dependence of the light energy and the number of excited atoms on time during the process involving the bleaching (opening) of a passive shutter: for $G > G_c$ a single pulse is generated (solid curve), while for $G < G_c$ relatively broad intensity peaks are generated (the dashed curve).

Thereafter, n continues to increase for a definite time, and at a certain time t_1 it reaches its maximum value $\bar{n} + \delta\bar{n}$ which slightly exceeds $\bar{n}$ (Fig. 44). Due to the stimulated emission $\delta\bar{n}$ from the excited atoms while the optical shutter is closed the latter receives an initial pulse, as a result of which it finally opens completely or partially, as was explained in § 45.

In order to find $\delta\bar{n}$ we make use of the fact that for $t < t_1$ the shutter is practically closed, and the time dependence of n is only slightly different from the dependence in the case of the conventional generation that occurs in a resonator with a fixed Q and is accompanied by relaxation oscillations (Fig. 27a). From this figure it is evident that the interval between the instant at which n is equal to the threshold value and the instant at which n reaches a maximum is close to a half-period of the oscillations (i.e., $t_1 = [2Y_1T/\nu\zeta\,(\varkappa_1 + \mu)]^{1/2}$; here ζ is the relative excess pump power above the threshold for a closed shutter, and $Y_1 \sim |\ln \delta\Omega| \gg 1$ is the initial amplitude of the relaxation oscillations, which is determined by Eq. (28.4)). We make use of Eq. (27.17) for the oscillation period after placing $Y = Y_1$ (this corresponds to the first period) and replacing $\varkappa_1$ by the quantity $\varkappa_1 + \mu_a$ (this corresponds to a closed shutter). During the time t_1 the quantity n increases by the following amount due to the pumping:

$$n(t_1) - \bar{n} \equiv \delta\bar{n} \cong t_1 \frac{N - N^*}{\hbar\omega_0} = \bar{n}\sqrt{\frac{2Y_1\zeta}{Tv(\varkappa_1 + \mu_a)}}. \tag{47.1}$$

By definition, $\dot{n} = 0$ at time t_1, and from Eq. (46.8) we have

$$J(t_1) = \frac{(N - N^*)}{vBn(t_1)}, \tag{47.2}$$

where N^* is the threshold pump power for a closed shutter.

Let us consider Eqs. (46.7), (46.8) in the range $t > t_1$ using the initial conditions (47.1), (47.2). When $t > t_1$ intense generation begins, and the deexciting action of the generated radiation sharply predominates over the effect of both the pumping and the spontaneous emission. Therefore, in the right side of Eq. (46.8) we need retain only the last term, and the equation takes the form

$$\hbar\omega_0\dot{n} = -vBnJ. \tag{47.3}$$

In view of the high energy density of the stimulated emission it is possible to neglect the spontaneous emission retained in the resonator in Eq. (46.7) and to drop the first term on the right side. On the other hand, as will be clear from what follows, the behavior of the optical shutter is already determined when it is slightly open (i.e., $\hbar\omega_0 n_a w_a / Jv\mu_a \gg 1$). Restricting ourselves to this range, we neglect unity in comparison with the quantity $\hbar\omega_0 n_a w_a / Jv\mu_a$ in Eq. (46.7). Dividing both sides of the equation by J and integrating with respect to time, we find

$$J = \frac{N - N^*}{vB\bar{n}} \exp\left\{vB\int_{t_1}^{t}[n(t) - \bar{n}]\,dt + \frac{v^2\mu_a^2}{\hbar\omega_0 n_a w_a}\int_{t_1}^{t} J dt\right\}. \tag{47.4}$$

with allowance for the initial conditions (47.1)-(47.2). Substituting J expressed in terms of n according to (47.3) into the exponent of this expression, we have

$$J = \frac{N - N^*}{vB\bar{n}} \exp\left\{-W + \frac{v\mu_a^2(\bar{n} + \delta\bar{n} - n)}{(\mu_a + \varkappa_1)n_a w_a}\right\}. \tag{47.5}$$

Here we have introduced the substitution

$$W = \frac{v(\mu_a + \varkappa_1)}{\bar{n}}\int_{t_1}^{t}[\bar{n} - n(t)]\,dt. \tag{47.6}$$

Substituting (47.5) into Eq. (47.3) and expressing n in terms of $\dot{W}$, we obtain an equation in W which can be integrated in quadratures. Taking the initial conditions (47.1) into account, we finally have [61]

$$\frac{t - t_1}{G} = \sqrt{\frac{T\underline{n}}{v\varkappa_1 \zeta \bar{n}}} \times$$

$$\times \int_0^{G\left[\sqrt{2Y_1} + \frac{\bar{n} - n}{\sqrt{\bar{n}\underline{n}}}\sqrt{\frac{v\varkappa_1 T}{\zeta}}\right]} \frac{dx}{G^2 + x + (e^x - 1)(G^2 + \sqrt{2Y_1}G - 1)} \tag{47.7}$$

Here†

$$G = \frac{\Delta\omega K_a \chi_a}{\Delta\omega_a} \cdot \frac{\sqrt{v\varkappa_1 T\zeta}\,(\bar{n} - \underline{n})}{\sqrt{\bar{n}\underline{n}}} \tag{47.8}$$

(we have expressed μ_a in terms of $\varkappa_1$ using Eq. (46.9)).

As was shown in § 45, the inertia of an optical shutter depends on the quantum luminescence yield K_a of the absorping medium. Therefore, the parameter G, which is proportional to K_a, characterizes the inertia of the optical shutter: as G increases, the inertia decreases.

The parameter G is represented as the product of two factors; the first of these is considerably less than unity, while the second is considerably greater than unity; therefore, G can vary over rather wide limits. The nature of the generation depends on whether or not G reaches its critical value

$$G_c = \sqrt{\frac{1}{2}Y_1 + 1} - \sqrt{\frac{1}{2}Y_1} \cong \frac{1}{\sqrt{2Y_1}}. \tag{47.9}$$

If $G > G_c$, then the coefficient of $(e^x - 1)$ in the denominator of the integrand in (47.7) is positive, while for $G < G_c$ this coefficient is negative. In the second case the integrand has a pole at the point x_0 which satisfies the equation

$$x_0 = \ln\left(1 + \frac{x_0 + G^2}{1 - G\sqrt{2Y_1} - G^2}\right) \cong \ln\left(1 + \frac{x_0}{1 - G/G_c}\right). \tag{47.10}$$

---

† For the case in which the absorbing centers function according to a multilevel diagram it is necessary to assume that the quantity A represents K_a in Eq. (47.8); here $1/T_a\tilde{w}_a$, where $\tilde{w}_a$ is the resultant probability (referred to a unit time) than an atom excited by the generated light in the absorbing medium will return to the ground state.

The quantity x_0 is positive; its graph is shown in Fig. 41. For $G < G_c$ the upper limit of the integral in (47.7) cannot exceed the point x_0 at which the integral diverges. But the value of the upper limit is determined by the number of excited active centers n, which therefore is bounded from below by a certain value n_{min}. A number of active centers not exceeding the value

$$\bar{n} - n_{min} = \left(\frac{x_0}{G} - \frac{1}{G_c}\right) \sqrt{\frac{\zeta \bar{n} n}{v \varkappa_1 T}}. \qquad (47.11)$$

can be deexcited in one peak. This number is an insignificant fraction (approximately 0.01) of the excess number of excited atoms $\bar{n} - \underline{n}$. Thus, for $G < G_c$ the optical shutter opens only slightly, and this is connected with its excessively high inertia.

Conversely, for $G > G_c$ (this corresponds to a fairly low inertia of the optical shutter) the upper limit of the integral in Eq. (47.7) can take any value, since the integrand does not have a pole; correspondingly, the entire reserve of excited atoms is deexcited in one peak (i.e., the optical shutter opens completely).

§ 48. The Different Operating Modes of a Laser with a Passive Shutter

1. An Operating Mode with a Single Pulse $G > G_c$. The optical shutter opens during a relatively long time $(2YT_{\underline{n}}/v\varkappa_1\zeta\bar{n})^{1/2}$, but the duration of the peak is a considerably smaller quantity of the order of $1/v\varkappa_1$ (Fig. 44). The intensity peak is practically no different in height and shape from the peak which develops in a resonator with an active shutter (§ 42); however, due to the considerably greater total duration of the generation the spectral width of the radiation is much smaller than it is in the case of an active shutter.†

In the case of a passive shutter the shape of the single pulse is described by the formulas in § 42, in which it is necessary to set $p = \bar{p} = \bar{n}/\underline{n}$.

2. The Operating Mode with Self-Oscillations Having Isolated Intensity Peaks. This operating mode is realized if the parameter G is slightly smaller than the critical

† Unlike the case of an active shutter, a relatively long duration of the generation (which is accompanied by a slow Q switching) does not lead to damping of the single pulse, since a passive shutter is maintained in the open state during the entire peak by the generated radiation.

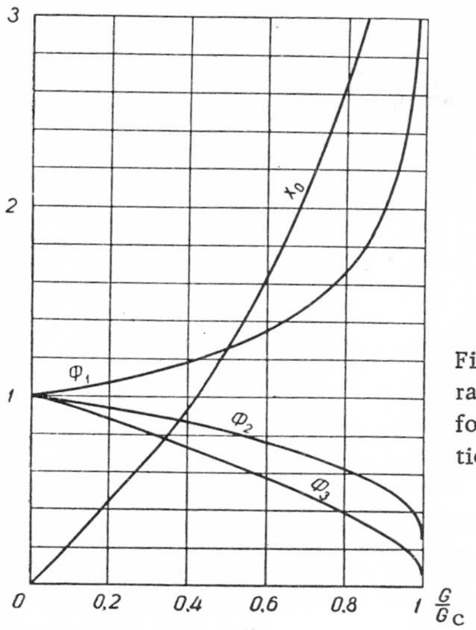

Fig. 45. Comparison of the peaks generated by a resonator with a passive shutter for G < G_c and for the case of free relaxation oscillations.

value G_c. The small number of excited atoms which are deexcited in one peak are rapidly replaced due to the pumping, and the peaks follow one another at equal intervals:

$$\Delta t = \sqrt{\frac{Tn}{v\varkappa_1\zeta\bar{n}}}\frac{x_0}{G} = \Delta t_{fr}\Phi_1\left(\frac{G}{G_c}\right) \tag{48.1}$$

The period Δt of the self-oscillations is represented as the product of the maximum period† of ordinary free oscillations in the presence of a closed shutter (§27) and the function $\Phi_1(G/G_c) = x_0G_c/2G$ (Fig. 45), which stipulates the ratio between the period of the intensity peaks in the presence of passive modulation and the period of the peaks of the free oscillations. For $G \to 0$ we have $\Delta t \to \Delta t_{fr}$, and for $G \to G_c$, $\Delta t \to \infty$. This is quite natural, since for T = 0 the optical shutter is constantly closed and there is no modulation, while for $G \geq G_c$ a single pulse is generated.

From Eq. (47.7), which determines the shape of the intensity peak, it is possible to find the half-width δt of the peak. Finally,

$$\delta t = 3.5\sqrt{\frac{Tn}{v\varkappa_1\zeta\bar{n}}}G_c\Phi_2\left(\frac{G}{G_c}\right) \equiv \delta t_{fr}\Phi_2\left(\frac{G}{G_c}\right), \tag{48.2}$$

† We recall that the period and amplitude of free relaxation oscillations decrease with time due to damping.

where $\delta t = \delta t_{fr}$ is the half-width of the peak for free oscillations having a maximum amplitude with the shutter closed; $\Phi_2(\xi) = \xi[-2\xi - 2\ln(1-\xi)]^{-1/2}$ is a function which stipulates the ratio between the width of the intensity peaks in the presence of passive modulation and the width of the peaks of the free oscillations (Fig. 45). For $G \to 0$, $\delta t = \delta t_{fr}$ and for $G \to G_c$, $\delta t \to 0$. The ratio between the half-width and the period of the self-oscillations has the form

$$\frac{\delta t}{\Delta t} = \left(\frac{\delta t}{\Delta t}\right)_{fr} \Phi_3\left(\frac{G}{G_c}\right), \tag{48.3}$$

where Φ_3 is the function shown in Fig. 45.

Equations (48.1)-(48.3) are valid only for isolated peaks. If the parameter G is close to G_c, then the period of the self-oscillations considerably exceeds the period of the free oscillations, so that it is possible to neglect interactions between peaks and to treat them as isolated peaks. However, if the difference $G_c - G$ is comparable with G_c, then the periods of the self-oscillations and the free oscillations do not differ by much, and in considering the peak it is necessary to take account of the stimulated-emission energy which has arrived from the previous peak. Therefore, for $G_c - G \sim G_c$ Eqs. (48.1)-(48.3) lose their validity, and the operating mode considered below occurs.

3. The Operating Mode with Self-Oscillations Having Mutually Coupled Intensity Peaks. Assume for simplicity that $G \ll G_c$. The consideration of interaction between peaks is expressed by the fact that the kinetic equations must be considered during the entire generation process rather than in the domain of a single peak; here it is necessary to consider both the pumping and the spontaneous emission. Using simple transformations which are similar to those described above, we shall reduce the kinetic equations (46.7), (46.8) to a form analogous to the equation for free relaxation oscillations (27.7):

$$y''(u) + \frac{dU}{dy} = Y\{\exp[-Y - y + Gy'(u)] - \exp[-Y - y]\}. \tag{48.4}$$

This equation differs from the equation for relaxation oscillations in the right side, which vanishes for $G \to 0$ (i.e., in the absence of modulation). For small G the right side of the equation can be treated as a small perturbing force associated with weak Q modulation. This force leads to a slow variation (in this case to an increase) of the amplitude Y of the oscillation. For a stipulated amplitude Y the shape and period of the oscillations are specified by the equations in § 27, in which it is necessary to replace $\varkappa_1$ by $\varkappa_1 + \mu_a$.

The law for the increase of the amplitude that accompanies an increase of the period number k is not difficult to establish; it has the form

$$k = \frac{1}{2} \int\limits_{G\sqrt{2Y_1}}^{G\sqrt{2Y}} \frac{x\,dx}{x\cosh x - \sinh x}. \tag{48.5}$$

From this it follows that until $G\sqrt{2Y} < 1$ the amplitude increases relatively slowly as follows:

$$Y_k \cong \left[\frac{1}{\sqrt{Y_1}} - \frac{\sqrt{8}}{3} Gk\right]^{-2}. \tag{48.6}$$

However, after Y reaches the value $1/2G^2$ the amplitude increases to infinity in one period, and the optical shutter would have had to be completely open. Actually, this does not occur, since the coupling between intensity peaks is broken before Y reaches the value $1/2G^2 \equiv Y(G_c/G)$, which considerably exceeds the maximum amplitude of the free oscillations.†

Actually, for $G \ll G_c$ generation does occur in this way. First, oscillations with the maximum amplitude Y_1 of the free oscillations develop. Because of Q modulation the value of Y increases somewhat during a small number of periods, after which the coupling between peaks is practically broken and the increase of the amplitudes ceases. Thus, generation is realized in the form of peaks which are somewhat narrower than the peaks of relaxation oscillations and are separated by a somewhat longer time interval.

The variation of the generation kinetics for a decrease of the parameter G is shown schematically in Fig. 46. For $G > G_c$ (Fig. 46a) a decrease of G has practically no effect on the shape of the single pulse. When G decreases to the value G_c, self-oscillations with a large period that decreases rapidly when G varies within the small range $0 < G_c - G \ll G_c$ develop instead of a single pulse (Fig. 46b). When G decreases so much that $G_c - G \sim G_c$, a

† If $G > G_c$ (i.e., $G\sqrt{2Y_1} < 1$), then the value of Y increases to infinity during just the first period (i.e., the optical shutter is completely open in accordance with what was said in §47).

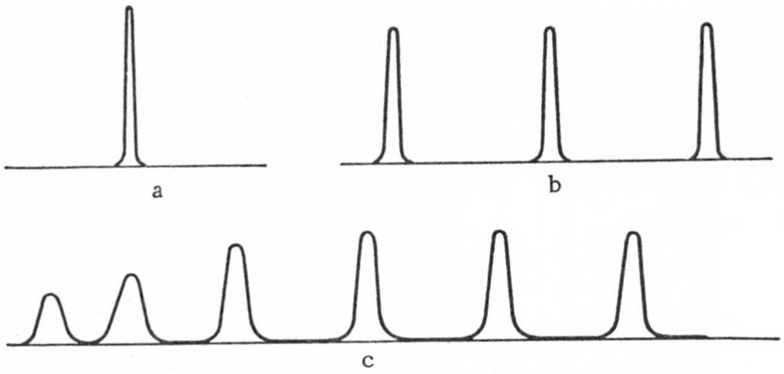

Fig. 46. Graph for the dependence of the generation kinetics on the parameter G: a) $G > G_c$, a single pulse; b) $0 < G_c - G \ll G_c$, self oscillations with isolated peaks; c) $G \ll G_c$, self-oscillations with mutually coupled peaks.

further decrease of G ceases to have a noticeable effect on the shape of the peaks, which remain somewhat sharper and less frequent than the peaks of the free oscillations (Fig. 46c). This continues until G becomes less than the value $(\xi^{1/2} + \xi^{-1/2})(v\varkappa_1 T)^{-1/2}$ that characterizes conventional damping of free oscillations; then the generation pattern ceases to differ from free relaxation oscillations.

One more generation process is possible besides the three enumerated above.

4. Practically Continuous Operation for Negative Q Modulation. This operating mode occurs when the action of the generated radiation causes the absorption of the passive shutter to increase rather than to decrease. Imagine, for example, that the absorbing medium has three levels which are approximately equally spaced: a ground-state level (1) and levels of excited states (2 and 3). Then the absorption of the generated light, which transfers the atoms from level 1 to level 2, will be accompanied not only by a decrease in absorption in the 1 → 2 band but also by the appearance of excited absorption in the 2 → 3 band. If the oscillator force for the 2 → 3 transition is greater than it is for the 1 → 2 transition, then the resultant absorption will increase under these conditions. Of course, other mechanisms which lead to a decrease in filter transparency due to the action of the generated light are also possible. Regardless of the specific mechanism, this leads to a change of the sign of the parameter G which characterizes

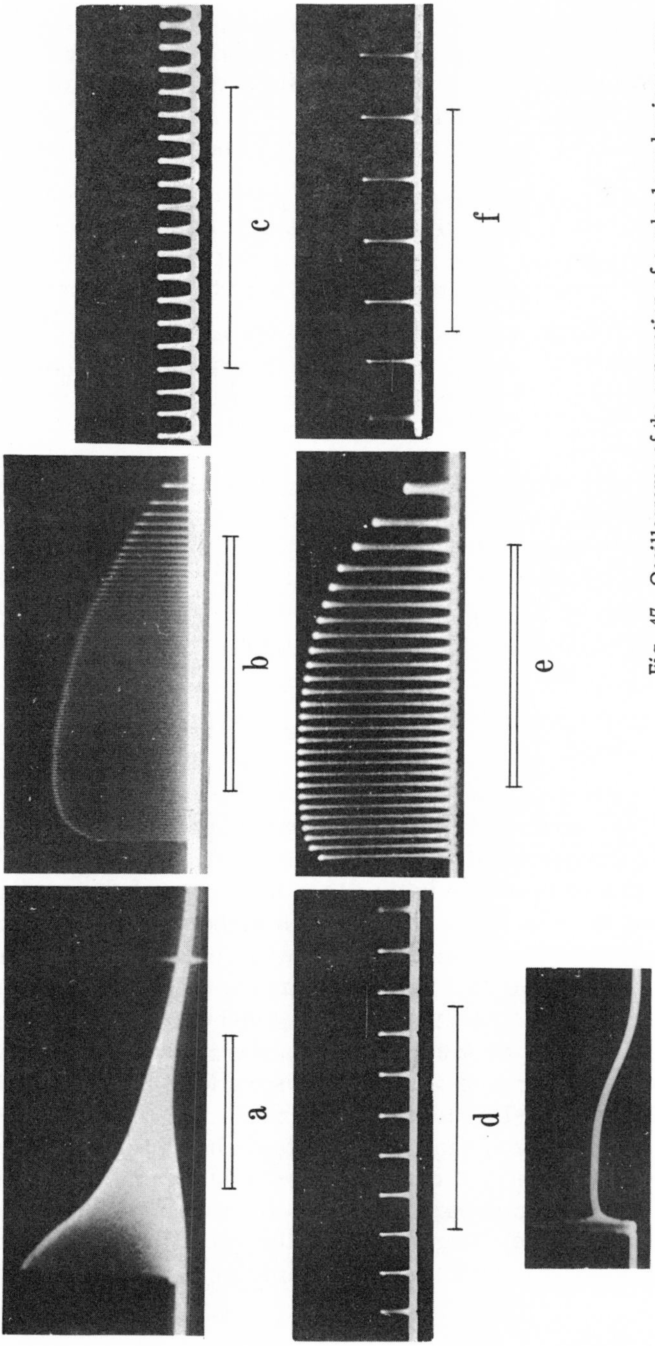

Fig. 47. Oscillograms of the generation of a ruby laser having a passive modulator in the form of a cryptocyanine solution (a-f) and a KS-15 filter with a negative modulating capacity (g).

the modulating capacity of the filter. For negative G the modulation leads to damping of the oscillations rather than to growth; this is evident, for example, from Eq. (48.6), which determines the damping law for $G < 0$. When the optical density of the modulator is not too small, this damping leads to practically continuous-wave generation.

The damping mechanism which accompanies negative modulation is immediately clear. The absorption of the filter increases at instants corresponding to the maximum intensity and decreases in the interval between peaks; this leads to a reduction of the peaks and to an increase of the intensity in the intervals.†

It should be noted that in order to obtain proper kinetic operation (self-oscillations or continuous-wave generation) it is necessary to satisfy the same conditions as those which are required for obtaining proper relaxation oscillations (§ 29). In particular, this can be achieved using a resonator having a sufficiently large angular divergence (for example, a resonator with lenses). This condition is not mandatory only when the parameter G is close to the critical value G_c or exceeds it (i.e., when generation is realized in the form of isolated peaks or a single pulse; however, in this case too the satisfaction of the above-mentioned conditions leads to a reduction of the duration of the peak – see §43).

The theory presented is in good agreement with experimental data [14, 116]. The parameter G can be changed by changing the quantum luminescence yield of the absorbing medium (for example, by changing the temperature), and also by changing its optical density (which is proportional to the difference $\overline{n} - \underline{n}$. A KS-15 filter was used as a negative modulator (the mechanism of the modulator action was not investigated). In all cases a change in G was accompanied by the described change in the kinetic operation. As an example, Fig. 47 shows several oscillograms [116] which were obtained for various concentrations of cryptocyanine: a) 0 (free generation); b, c) 10^{-5} g/liter (self-oscillations with mutually coupled peaks); d) 10^{-4} g/liter; e, f) $5 \cdot 10^{-4}$ g/liter (self-oscillations with isolated peaks); g) KS-15 filter.

An ordinary line indicates an interval of 20 μsec; a double line indicates an interval of 200 μsec.

† The negative feedback between the generation intensity and the Q of the resonator, which was simulated by means of a Kerr cell, leads to continuous-wave operation [104].

Conclusions

1. Q modulation by means of a passive optical shutter leads to generation either in the form of a single pulse or in the form of self-oscillations, depending on whether the parameter G characterizing the inertia of the absorbing medium reaches its critical value G_c. If $G > G_c$, a single pulse is generated; if G is somewhat less than G_c, self-oscillations with a relatively long period and sharp peaks occur; if $0 < G_c - G \lessapprox G_c$, then the period of the self-oscillations does not exceed the period of ordinary free relaxation oscillations by very much.

2. Passive Q modulation for $G > G_c$ can be compared with active one-shot modulation, while passive modulation for $G < G_c$ can be compared with active periodic Q-modulation. The difference consists in the fact that for $G > G_c$ the spectral width of a single pulse is greater in the case of active modulation than it is in the case of passive modulation, while for $G < G_c$ passive modulation is accompanied by less frequent and sharper peaks than active modulation, which is in resonance with strong free oscillations (§44).

3. The use of a filter having a negative modulating capacity leads to continuous-wave generation.

4. In order to obtain proper self-oscillations for $G \ll G_c$ it is necessary to satisfy the same conditions as those required for obtaining proper free relaxation oscillations (§ 29); these conditions are satisfied, for example, in the case of a resonator with positive lenses or concave reflectors.

Chapter XII

MODE LOCKING AND ULTRASHORT PULSES

Richard A. Phillips

In this section the influence of the relative phases of modes on the output waveform and on the operation of a laser will be considered. Previously it was shown that the operating frequencies of a laser are to first order determined by the cavity resonances and longitudinal modes (TEM_{00q}) are separated in frequency by $\Delta\nu = c/2Ln$, where L is the cavity length and n is the index of refraction of the medium. We will see that when the phases of the modes are locked the output consists of a series of pulses. In an absorbing medium the pulse velocity has been measured as greater than c. From solid-state lasers pulses on the order of 10^{-11} sec have been observed.

§ 49. The Electric Field of a Multimode Laser

The electric field E at the output mirror of a laser is equal to the sum of the fields from individual modes. The time-dependent field E(t) at this point is given by

$$E(t) = \sum_m E_{om} \cos(\omega_m t + \varphi_m) , \tag{49.1}$$

where ω_m is the frequency, E_{om} the amplitude, and φ_m the phase of the field from the m-th mode. Ignoring for the moment the effects of the active media, ω_m is given by the expression

$$\omega_m = \omega_o + m\Delta\omega , \tag{49.2}$$

where ω_0 is the frequency of the central mode. The frequency of the m-th mode is $\omega_m + \varphi_m \cdot \varphi_m$ is a lowly varying function of time in most lasers. For a short time it can be considered a constant. When the output consists of several equally spaced modes, the total field E(t) and the intensity $E^2(t)$ are periodic functions of time with period $\Delta\nu^{-1}$. The relative phases of the modes are important since they determine the manner in which the individual modes combine.

When the phases of the modes are unlocked, i.e., random, the field from each mode reaches its maximum at a different instant. Consequently, the variation of E^2 with time is of the type shown in Fig. 48a. In an unlocked laser mode pulling, mode competition, and phase changes produce variations in the frequencies of the modes. Under these conditions the modes are not equally spaced and the periodicity of the output cannot be sustained. Such a laser is noisy particularly in the kilocycle region and below.

On the other hand, when the phases are locked, i.e., all equal, the field from each mode reaches a maximum at the same instant. For example, if $\varphi_m = 0$ for each m, then the field of each mode is a maximum at t = 0, Fig. 48b. In a locked laser pulling and competition effects are eliminated and the modes are precisely equally spaced in frequency. In He — Ne lasers, the noise is reduced about five orders of magnitude from that in the unlocked condition.

For the same average power, the unlocked and locked outputs, Fig. 48a and 48b, have different peak powers. For modes of equal amplitude the ratio of the peak power of the mode locked to unlocked case is approximately equal to the number of modes. To show this, let us assume that there are M modes. When their phases are locked, the maximum E is ME_{om}; the peak power is proportional to

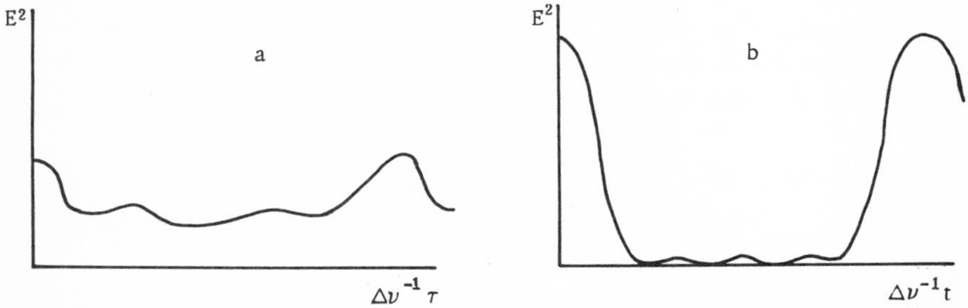

Fig. 48. The ouput power from a laser is proportional to the square of the sum of the fields in the individual modes. For equally spaced modes the ouput is periodic with period $\Delta\nu^{-1}$. In (a) the phases of the modes are random. In (b) the phases are locked and the output consists of narrow pulses.

$M^2E^2_{om}$ and the average power is proportional to ME^2_{om}. When their phases are random, the peak power and average power are both proportional to ME^2_{om}. Thus the ratio of the average powers is unity, whereas the ratio of the peak powers is M.

Therefore, the output of a mode locked laser consists of a series of pulses. The greater the number of modes, the narrower the pulse width. It should be emphasized that this type of pulsed output is distinct from the relaxation oscillations described in earlier sections. (The period of relaxation oscillations is related to the relaxation time of the upper level of the laser transition whereas the period of mode-locked pulses is equal to the round trip time for light in the laser cavity, $\Delta\nu^{-1}$. Both periods are different from the period of Q switched pulses when a rotating prism is used on a Q switching unit. The latter period is determined by the rotational speed of the prism.)

When the output of a laser, observed at one mirror, consists of a series of pulses separated by $t = 2nL/c$, then inside of the cavity there must be a pulse which travels from one mirror to the other. Each time the pulse arrives at the output mirror, a fraction of it is transmitted. This will be shown below.

The electric field E in a laser cavity can be represented as the sum of the fields in each cavity mode. In a standing wave system

$$E(z,\ t) = \sum_m E_{om}(e^{ik_m z} - e^{-ik_m z})e^{i\omega_m t}\,, \tag{49.3}$$

where $z = 0$ at one mirror and $z = L$ at the other. k_mL is determined from the boundary condition $k_mL = 2\pi q$. The central mode of the system is designated by $m = 0$, k_0, and ω_0. Adjacent modes have positive and negative values of the integer m. The quantities ω_m and k_m are then given by

$$\omega_m = \omega_o + m\Delta\omega\,, \tag{49.2}$$

$$k_m = k_o - m\pi/L\,. \tag{49.4}$$

Adding the fields for $2M + 1$ modes yields

$$E = e^{i(\omega_o t - k_o z + \pi/2)} \sum_{m=-M}^{M} E_{om} e^{im(\Delta\omega t - \pi z/L)} + e^{i(\omega_o t + k_o z - \pi/2)} \sum_{m=-M}^{M} E_{om} e^{im(\Delta\omega t + \pi z/L)}\,. \tag{49.5}$$

The terms in front of the summation signs describe right and left traveling waves with average frequency ω_0. This is the carrier frequency. The summation terms vary slowly with t and z and are envelope terms for the two carriers. When $t = 0$ and $z = 0$ these envelope terms are a maximum; they add to give a pulse. At a later time, $t = \pi/\Delta\omega$, these envelope terms are a maximum at $z = L$, i.e., at the other mirror. We conclude that a pulse travels back and forth between mirrors inside of the cavity. The pulse velocity is $L\Delta\omega/\pi$.

Another type of mode locking takes place when the even numbered modes are locked in phase with $\varphi = 0$ and the odd numbered modes are locked in phase with $\varphi = \pi$. In this case two pulses travel in the cavity. The output waveform consists of pulses separated by half the period of the previous case. The two pulses which make up this output each are separated by $2L/c$. One pulse is displaced by L/c with respect to the other one, and one is out of phase with respect to the other by π. Interference experiments have confirmed that the relative phase is π.

Under certain conditions every second mode can be suppressed. The output in this case also consists of pulses spaced by L/c.

We will now describe the experimental conditions for mode locking.

§ 50. Forced Mode Locking

The simplest method of obtaining mode locking is to insert a variable loss into the cavity near one mirror [146–148, 161]. The loss is made equal to zero at intervals equal to the round trip time for a wave inside the cavity, i.e., $t = 2L/c$. When this is done the laser is forced to operate in a pulsed manner; this in turn requires mode locking.

A variable loss can be provided by an electrooptic cell, using either the linear (Pockels) [146] or quadratic (Kerr) electrooptic effect, combined with two polarizers. (If there is a surface at Brewsters angle inside the cavity, the polarizers can be dispensed with.) The transmission through this system depends on the voltage applied to the cell. When a sinusoidal voltage with period L/c is applied to the cell, the combination will have complete transmission at intervals of $L/2c$.

The linear electrooptic effect is described by the relation

$$P^{\omega} = XE^{\omega}E^{0}, \tag{50.1}$$

where P is the polarization and X is the susceptibility for the linear electrooptic effect (actually a tensor). The superscripts denote the frequencies of the fields.

The polarization of the wave entering the cell is set at 45° with respect to the axis induced by the applied voltage. Let $\overline{E}_{in}$ be the incoming amplitude of the wave. Then it is decomposed into two perpendicular eigenmodes of the cell, $\overline{E}_1$ and $\overline{E}_2$:

$$\overline{E}_{in} = \overline{E}_1 + \overline{E}_2 . \tag{50.2}$$

Upon emerging from the cell a relative shift δ is introduced for one wave. The phase shift for the linear electrooptic effect is given by

$$\delta = \frac{\pi l}{\lambda n} XE^2, \tag{50.3}$$

where l is the length of the crystal; n is the average index of refraction for the two eigenmodes. We then have

$$\overline{E}_{out} = \overline{E}_1 + \overline{E}_2 e^{i\delta} . \tag{50.4}$$

In general the emerging wave is elliptically polarized. When $\delta = \pi/2$ it is circularly polarized; when $\delta = \pi$ it is linearly polarized but the plane of polarization makes an angle of 90° with respect to the incoming plane of polarization. The intensity transmitted through the analyzer varies as $\cos^2(\delta/2)$.

An electrooptic crystal driven at $\nu = L/c$ can be force mode locking in another way, i.e., by generating combination frequencies which then act as injection signals [155, 159]. Combination frequencies arise from the same nonlinear polarization term as in Eq. (50.1):

$$P^{\omega \pm \Delta\omega} = XE^{\omega}E^{\Delta\omega} . \tag{50.5}$$

The time varying polarization at frequency $\omega \pm \Delta\omega$ generates a wave at this frequency. The phase of this generated wave is locked to the phases of the applied waves at ω and $\Delta\omega$.

Another and more economical way of attaining mode locking is to insert a saturable absorber into the cavity near one mirror [149–150]. The absorption coefficient depends on the population difference between molecules in the upper and lower levels of the absorbing transition. For a given number of centers a sufficiently intense beam of light can saturate the absorption. Saturation occurs when the number of molecules in the upper and lower levels is

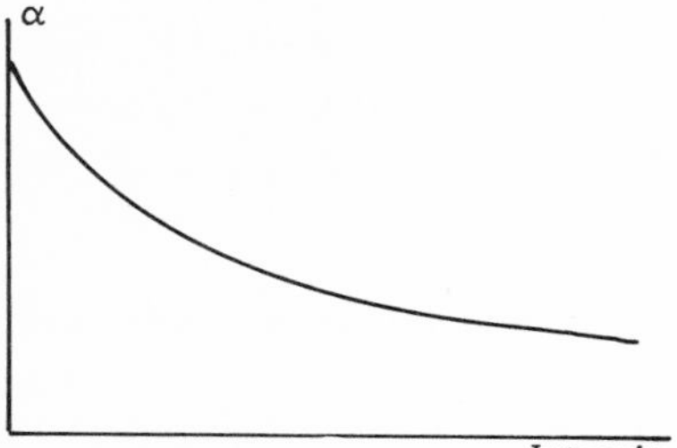

Fig. 49. The absorption of a sample is proportional to the population difference between the lower and upper states ($N_1 - N_2$). This difference depends on the intensity of light passing through the sample.

equal; in this state the sample becomes transparent, Fig. 49. When the proper values are chosen for the relaxation time, intensity of light, number of absorbing centers, and their cross section, the absorber opens at intervals of $\Delta\nu^{-1}$. This kind of operation requires that the modes be locked. As the periodic wave passes through this nonlinear absorber, the low-intensity portion of the wave is absorbed, but the peaks saturate the absorber and are transmitted. The peaks are also narrowed in the process. Narrowing to $\Delta\nu^{-1}$ produces mode locking and enhances the peak power. Then further narrowing of the peak brings new modes into oscillation, increasing the bandwidth of the laser output (Figs. 50 and 51). According to well known results from communication theory, a decrease in the pulse width must be accompanied by an increase in its bandwidth. In Nd^{+3} glass lasers the bandwidth is on the order of 100^{-1} cm and the pulse width is on the order of 10^{-11} sec.

The detection and observation of these ultrashort pulses [150] is an accomplishment in its own right. It should be emphasized that a train of pulses is produced rather than a single pulse. When such a train undergoes reflection by a mirror and retraces its path, standing waves are set up. Antinodes exist at spacings of L/c. They were detected in [150] by inserting a liquid into the standing wave region. The liquid was transparent at the laser frequency but strongly absorbing at twice the laser frequency. At the antinode the photon density was so high that two-photon absorption took place. The excited molecules in the material then decayed to the ground state via fluorescence. A photograph was taken of the laser path and the width of the fluorescing region measured. This width corresponded to the pulse width.

§ 51. Self-Mode-Locking

Spontaneous mode locking occurs for a suitable choice of laser operating parameters [152, 153]. The theory of self-mode-locking is too involved to be presented here in complete form. Instead the main feature will be outlined. There are two approaches used in describing self-mode-locking: one in the time domain and the other in the frequency domain.

In the frequency domain approach, due to Lamb [154], two waves at ω and $\omega + \Delta\omega$ are mixed by the third-order nonlinear polarization to give an injection signal at $\omega - \Delta\omega$. The polarization third order in the electric field is described by the expression

$$P^{\omega-\Delta} = X^3 E^{\omega} E^{\omega} E^{-(\omega+\Delta\omega)}, \tag{51.1}$$

where X^3 is the third-order nonlinear susceptibility. Its tensor properties can be suppressed for the discussion that follows. The nonlinear polarization equation (51.1) acts as a source term in the wave equation and generates a wave at this frequency. These combination frequencies were observed directly in [162]. This injected wave is amplified by the active medium and the mode at $\omega - \Delta\omega$ has its phase locked to the modes at ω and $\omega + \Delta\omega$. Modes on either side of the locked set are then in turn locked to them.

In Lambs' theory, the injection signal which produces mode-locking has to be strong enough to overcome a competing effect: mode-pulling. When an active medium is in the cavity, the mode frequencies are pulled toward the center of the atomic resonance. The amount of pulling depends on the gain and the difference between the frequency of the mode and the line center. The injection signal has to be strong enough so that the laser amplifies it rather than operating on the preferred (pulled) frequency.

Since the pulling effects are known, the calculation of the conditions under which self-mode-locking takes place requires a knowledge of the strength of the injection signal. This in turn requires that the numerical value of X^3 be known. It is difficult to

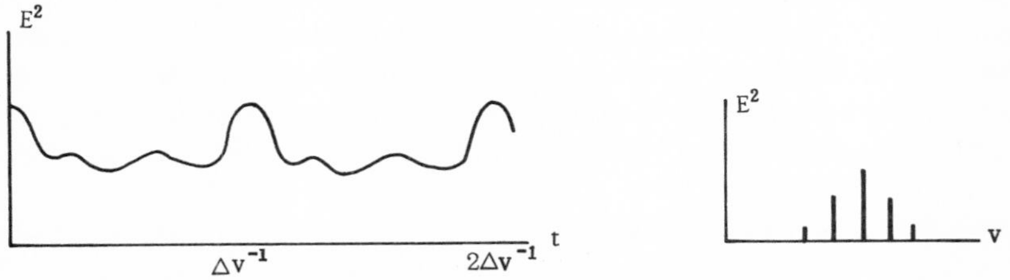

Fig. 50. The output of a multimode unlocked laser is periodic. The variations in intensity during the period are small.

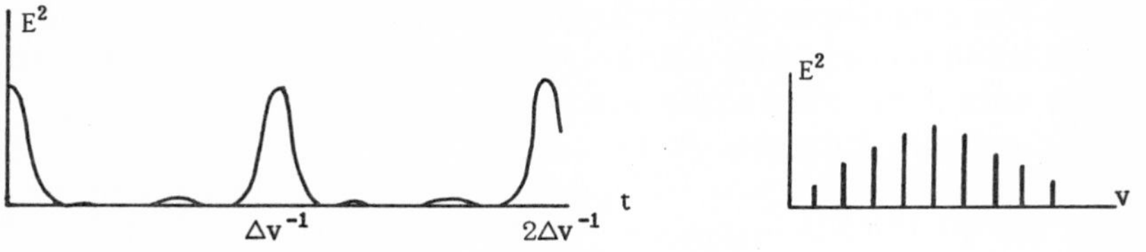

Fig. 51. When the wave shown in Fig. 50 passes through a saturable absorber the high-intensity peaks are transmitted but the low-intensity portions are absorbed. This sharpens the peaks, resulting in mode-locking and increased bandwidth.

calculate the quantity quantum-mechanically. Nash [155] obtained a value for X^3 by indirect means. He used a KDP crystal inside the cavity to generate an injection signal. Measuring the rf voltage that produced marginal locking and knowing the values of X for KDP (3×10^{-7} e.s.u.), he was able to calculate the strength of the injection signal. Then he adjusted the laser for self-locking and measured the field strengths inside the cavity. From these data he calculated $X^3 = 3 \times 10^{-8}$ e.s.u. This was almost as large as the value of X for KDP because of resonant enhancement of the coefficient.

In the time domain approach the growth and sharpening of the pulse is produced by the time-dependent behavior of the gain. Initially the phases are unlocked but there are small periodic peaks in the intensity. When the wave is amplified the increase in intensity, dI, is proportional to the population inversion $(N_2 - N_1)$ times the intensity I:

$$dI \propto (N_2 - N_1)I \; . \tag{51.2}$$

Consequently when the initially unlocked waveform passes through the amplifying region, the peaks receive more energy. This reduces the inversion and also reduces the gain for the part of the wave following the peaks. This process enhances the peaks and produces a pulsed output. The time for the gain to recover is on the order of the homogeneously broadened lifetime. One would therefore expect that self-mode-locking would take place most easily when the round trip time was comparable to this lifetime. Using this approach one might expect that a laser would always self-lock, and this does not happen. When the modes are not equally spaced, the periodic output quickly dies out. The peaks have to grow at a fast enough rate to overcome effects of phase fluctuations and mode-pulling.

The conditions under which self-locking of He−Ne lasers has been observed [156] will be described. Figure 52 shows the power output vs. single trip loss for a wave in a 110 cm cavity having a 2 mm diameter 81 cm long discharge tube filled with 2.2 torr He and 0.2 torr Ne. The power output reaches a maximum when the loss is approximately 2.5%. It was found experimentally by Faxvog [156] that when the output coupling was slightly in excess of 2.5% the laser spontaneously became mode-locked. In the experiment a thin plate of fused silica was inserted into the cavity so that the angle between the normal of the plate and the laser beam was initially at Brewsters angle. The plate was then rotated while the total power coupled out of the cavity was measured. The output power consisted of a beam transmitted by one mirror plus the beams reflected by the fused silica plate. The length of the cavity, 110 cm, gave a round trip time of 7×10^{-9} sec. When the cavity was lengthened, pulses appeared at intervals of L/c. The interpretation of this behavior was that two of the pulses were sustained; every other mode was phase-locked with $\phi = 0$ whereas the intermediate modes were locked with $\phi = \pi$.

§52. Pulse Velocity Inside the Cavity

A very interesting question can be raised in connection with mode-locking: how does the pulse propagate in a medium without being distorted. From the periodic output it is concluded that the pulse in the cavity retains its shape. Yet it is well known that when a narrow pulse propagates in a dispersive medium each Fourier component of the pulse propagates with slightly different velocity.

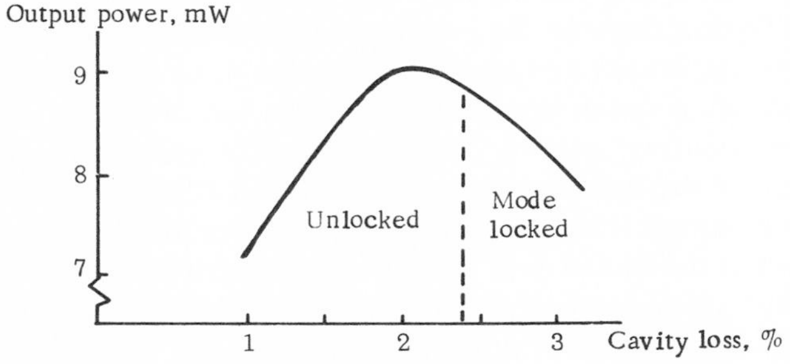

Fig. 52. The output power vs. cavity loss for a 1.1 m He−Ne laser. When the loss exceeded 2.4% stable spontaneous mode-locking was observed.

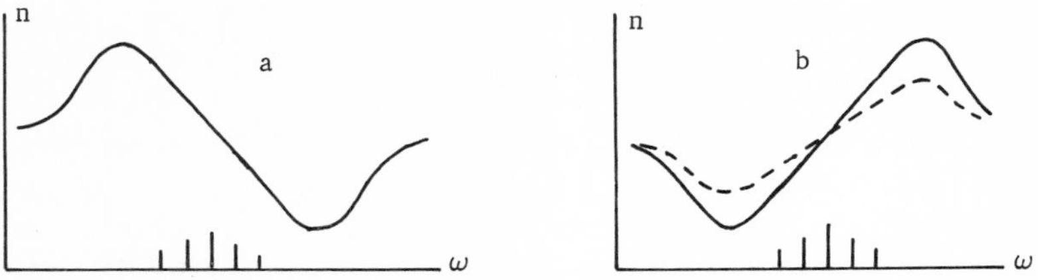

Fig. 53. The dispersion curve for an absorbing (a) and amplifying (b) medium. In (b) the dashed curve represents the condition that the population inversion is smaller.

The various components get out of phase and they no longer peak simultaneously. This causes the pulse to spread out.

The crest of a pulse propagates with group velocity as given by the expression [157–158]

$$\frac{d\omega}{dk} = \frac{c}{n + \omega \dfrac{dn}{d\omega}} \quad . \tag{52.1}$$

The frequency of a laser pulse corresponds approximately to the center of the laser transition. This is the region of anomalous dispersion. For frequencies close to the center of the line the variation of n with ω is approximately linear, Fig. 53. In an absorbing material $dn/d\omega$ is negative and the quantity $\omega(dn/d\omega)$ can be sufficiently large so that the group velocity is greater than c, the velocity of light in a vacuum. Pulse velocity in a Ne absorption cell was measured to be greater than c by a few parts in 10^4 by Faxvog, Chow, Bieber, and Carruthers [156]. This does not violate the principle that energy cannot travel faster than the speed of light. What happens is that absorption takes place at the trailing edge of the pulse (it does not take place instantaneously when a

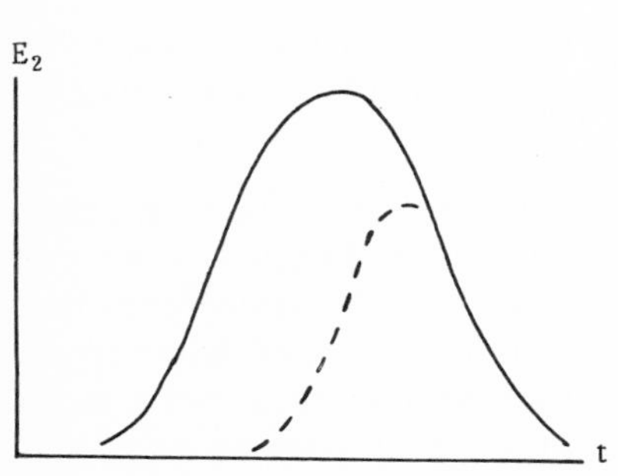

Fig. 54. Absorption at the trailing edge of a pulse causes the peak to be dispaced forward.

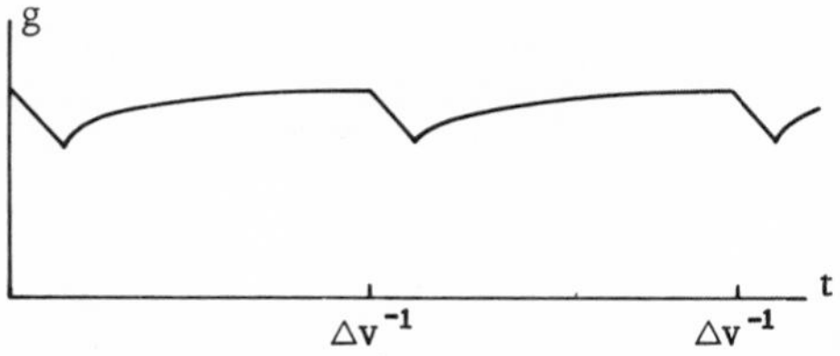

Fig. 55. The gain vs time for a laser medium operating in the pulsed conditions.

wave is incident on an atom but rather requires several cycles to develop). The effect is to move the peak forward. Energy is propagating at less than c but the pulse maximum is propagating at a velocity greater than c.

In a gain medium, on the other hand, the dispersion curve is reversed because the population is inverted, Fig. 53. $\omega\, dn/d\omega$ is positive and the pulse travels slower than c. This was verified in [158].

Why doesn't the pulse spread out? Because the inversion changes as the pulse travels through the medium [160]. This reduces the slope of the dispersion $dn/d\omega$ at the center of the line as shown in the dotted curve of Fig. 53. Therefore the trailing edge tends to move slightly faster than the leading edge. This compresses the pulse counteracting the effect of dispersion which spreads it out. The frequency of the pulse is chirped by this process; it is different at the leading than at the trailing edge.

Finally, it must be mentioned that the above discussion applies to a laser undergoing small variations of gain with time and having relatively low light intensity in the cavity. In the other limit the intensity is strong enough that the pulse greatly modifies the gain. In a π-pulse the pulse width is shorter than the time necessary to attain thermodynamic equilibrium and the pulse intensity is so strong that the population inversion is reversed during its passage.

We have seen that longitudinal modes give rise to a pulse which travels back and forth along the axis inside of a laser cavity. In a similar manner it can be shown that the transverse modes give rise to motion of the pulse in the place perpendicular to the axis. The combination of both longitudinal and transverse modes gives a packet or ball of energy which rapidly bounces back and forth along the axis while slowly sweeping out the cross section of the laser medium!

Conclusions

1. When the phases of the longitudinal modes are locked, i.e., equal, the output of the laser consists of periodic pulses.

2. A multimode laser can be forced to operate in the phase-locked condition by inserting a time varying loss into the cavity. The loss should be zero at intervals corresponding to the round trip time for light in the cavity.

3. Under suitable operating conditions self-locking occurs.

4. Pulse velocity greater than c, the velocity of light in free space has been measured in a Ne absorption cell inside of the cavity of a self-locked He–Ne laser.

5. Pulse width is limited by the relation $\tau > \Delta\nu^{-1}$. In ruby and Nd^{+3} lasers, pulse widths on the order of 10^{-11} sec have been observed.

REFERENCES

1. R. V. Ambartsumyan, N. G. Basov, P. G. Kryukov, and V. S. Letokhov, Zh. Éksper. i Teoret. Fiz. (Pis'ma), 3:261 (1966).
2. R. V. Ambartsumyan, N. G. Basov, P. G. Kryukov, and V. S. Letokhov, Zh. Éksper. i Teoret. Fiz., 50:724 (1966).
3. R. V. Ambartsumyan, N. G. Basov, P. G. Kryukov, and V. S. Letokhov, Zh. Éksper. i Teoret. Fiz., 51:1669 (1966).
4. Yu. A. Anan'ev, Zh. Tekh. Fiz., 37:139 (1967).
5. Yu. A. Anan'ev, A. A. Mak, and B. M. Sedov, Zh. Éksper. i Teoret. Fiz., 52:12 (1967).
6. Yu. A. Anan'ev and B. M. Sedov, Zh. Éksper. i Teoret. Fiz., 48:779 (1965).
7. N. G. Basov, R. V. Ambartsumyan, V. S. Zuev, P. G. Kryukov, and V. A. Letokhov, Zh. Éksper. i Teoret. Fiz., 50:23 (1966).
8. N. G. Basov, R. V. Ambartsumyan, V. S. Zuev, P. G. Kryukov, and V. S. Letokhov, Doklady Akad. Nauk SSSR, 165:58 (1965).
9. N. G. Basov, V. S. Zuev, and Yu. V. Senatskii, Zh. Éksper. i Teoret. Fiz. (Pis'ma), 2:57 (1965).
10. N. G. Basov and V. S. Letokhov, Doklady Akad. Nauk SSSR, 167:73 (1966).
11. N. G. Basov, V. N. Morozov, and A. N. Oraevskii, Doklady Akad. Nauk SSSR, 162:781 (1965).
12. N. G. Basov, V. N. Morozov, and A. N. Oraevskii, Zh. Éksper. i Teoret. Fiz., 49:895 (1965).
13. N. K. Bel'skii and D. A. Mukhamedova, Doklady Akad. Nauk SSSR, 158:317 (1964).
14. É. G. Berzing and Yu. V. Naboikin, Zhur. Priklad. Spektr., 5:31 (1966).
15. B. L. Borovich, V. S. Zuev, and V. A. Shcheglov, Zh. Éksper. i Teoret. Fiz., 49:1031 (1965).
16. S. L. Borovitskii, Yu. M. Gryaznov, and A. A. Chastov, Zh. Priklad. Spectr., 5:609 (1966).
17. V. L. Broude, V. I. Kravchenko, N. F. Prokopyuk, and M. S. Soskin, Zh. Éksper. i Teoret. Fiz. (Pis'ma), 2:519 (1965).
18. V. L. Broude, V. S. Mashkevich, A. F. Prikhot'ko, N. F. Prokopyuk, and M. S. Soskin, Fiz. Tverd. Tela, 4:2980 (1962).
19. V. L. Broude, O. N. Pogorelyi, and M. S. Soskin, Doklady Akad. Nauk SSSR, 163:1342 (1965).

20. V. L. Broude and M. S. Soskin, in: Quantum Electronics, 1, Naukova Dumka, Kiev (1966), p. 123.
21. G. Yu. Buryakovskii and V. S. Mashkevich, Ukrainskii Fiz. Zh., 10:65 (1965).
22. L. A. Vainshtein, Zh. Éksper. i Teoret. Fiz., 44:1050 (1963).
23. L. A. Vainshtein, Zh. Éksper. i Teoret. Fiz., 45:684 (1963).
24. M. P. Vanyukov, V. I. Isaenko, L. A. Luizova, and O. A. Shorokhov, Zh. Priklad. Spektr., 2:295 (1965).
25. M. P. Vanyukov, V. I. Isaenko, and V. V. Lyubimov, Zh. Priklad. Spektr., 3:171 (1965).
26. M. P. Vanyukov, V. I. Isaenko, and V. A. Serebryakov, Zh. Éksper. i Teoret. Fiz., 47:2019 (1964).
27. T. V. Gvaladze, I. K. Krasyuk, P. P. Pashinin, A. V. Prokhindeev, and A. M. Prokhorov, Zh. Éksper. i Teoret. Fiz., 48:106 (1965).
28. M. E. Globus, Yu. V. Naboikin, A. M. Ratner, I. A. Rom-Krichevskaya, and Yu. A. Tiunov, Zh. Éksper. i Teoret. Fiz., 52:859 (1967).
29. G. O. Karapetyan, Ya. É. Kariss, S. G. Lunter, and P. P. Feofilov, Zh. Priklad. Spektr., 1:93 (1964).
30. V. V. Korobkin and A. M. Leontovich, Zh. Éksper. i Teoret. Fiz., 44:1847 (1963); 49:10 (1965).
31. V. V. Korobkin, A. M. Leontovich, and M. N. Smirnova, Zh. Éksper. i Teoret. Fiz., 48:79 (1965).
32. N. G. Kramarenko, A. V. Meshcheryakov, Yu. V. Naboikin, A. M. Ratner, and I. A. Rom-Krichevskaya, in: Quantum Electronics, 1 [in Russian], Naukova Dumka, Kiev, (1966), p. 144.
33. M. A. Krivoglaz, Zh. Éksper. i Teoret. Fiz., 25:191 (1953).
34. M. A. Krivoglaz and S. I. Pekar, Tr. Inst. Fiz. Akad. Nauk Ukr. SSR, No. 4, p. 37 (1953).
35. A. M. Kubarev and V. I. Piskarev, Zh. Éksper. i Teoret. Fiz., 48:1233 (1965).
36. L. D. Landau and E. M. Lifshits, Electrodynamics of Continuous Media [in Russian], GITTL, Moscow (1967).
37. L. D. Landau and E. M. Lifshits, Mechanics [in Russian], Nauka, Moscow (1965).
38. O. L. Lebedev, V. N. Gavrilov, Yu. M. Gryaznov, and A. A. Chastov, Zh. Eksper. i Teoret. Fiz. (Pis'ma), 1(2):14 (1965).
39. V. L. Levshin, Zh. Fiz. Khim., 2:641 (1931).
40. B. Lindell, Lasers [Russian translation], Mir, Moscow (1964).
41. V. S. Letokhov and A. F. Suchkov, Zh. Éksper. i Teoret. Fiz., 50:1148 (1966).
42. B. L. Livshits and V. N. Tsikunov, Doklady Akad. Nauk SSSR, 162:314 (1965).
43. B. L. Livshits and V. N. Tsikunov, Doklady Akad. Nauk SSSR, 163:870 (1965).
44. B. L. Livshits and V. N. Tsikunov, Zh. Éksper. i Teoret. Fiz., 49:1843 (1965).
45. B. L. Livshits and V. N. Tsikunov, Ukrainskii Fiz. Zhur., 10:1267 (1965).
46. B. L. Livshits, V. P. Nazarov, L. K. Sidorenko, and V. N. Tsikunov, Zh. Éksper. i Teoret. Fiz. (Pis'ma), 1(5):23 (1965).
47. Ch. B. Lushchik, N. E. Lushchik, and K. K. Shvarts, Tr. Inst. Fiz. i Astron. Akad. Nauk Ést. SSR, 8:3 (1968).
48. V. I. Malyshev, A. S. Markin, and V. S. Petrov, Zh. Priklad. Spektr., 3:415 (1965).
49. V. I. Malyshev, A. S. Markin, and V. S. Petrov, Zh. Éksper. i Teoret. Fiz. (Pis'ma), 1:49 (1965).

50. V. S. Mashkevich, Ukrainskii Fiz. Zh., 10:55 (1965).
51. V. S. Mashkevich, Fundamentals of Laser-Radiation Kinetics, , Naukova Dumka, Kiev (1966).
52. A. L. Mikaélyan, A. V. Korovitsyn, and A. V. Naumova, Zh. Éksper. i Teoret. Fiz. (Pis'ma), 2:37 (1965).
53. A. L. Mikaélyan and Yu. G. Turkov, Radiotekh. i Élektron., 9:743 (1964).
54. Yu. V. Naboikin, A. M. Ratner, I. A. Rom-Krichevskaya, and Yu. A. Tiunov, in: Quantum Electronics, 2, Naukova Dumka, Kiev (1967), p. 146.
55. S. I. Pekar, Zh. Éksper. i Teoret. Fiz., 22:641 (1952).
56. A. M. Prokhorov, Radiotekh. i Élektron., 8:1073 (1963).
57. A. M. Ratner, Fiz. Tverd. Tela, 3:704 (1961).
58. A. M. Ratner, Fiz. Tverd. Tela, 1:1907 (1959).
59. A. M. Ratner, Zh. Tekh. Fiz., 34:115 (1964).
60. A. M. Ratner, in: Quantum Electronics, 1, Naukova Dumka, Kiev (1966), p. 150.
61. A. M. Ratner, in: Quantum Electronics, 1, Naukova Dumka, Kiev (1966), p. 166.
62. A. M. Ratner, in: Quantum Electronics, 2, Naukova Dumka, Kiev (1967), p. 75.
63. A. M. Ratner, Ibid., p. 53.
64. A. M. Ratner, Ibid., p. 91.
65. A. M. Ratner, Opt. i Spektr., 18:258 (1965).
66. A. M. Ratner, Zh. Éksper. i Teoret. Fiz., 45:1908 (1963).
67. A. M. Ratner, Zh. Éksper. i Teoret. Fiz., 52:1745 (1967).
68. A. M. Ratner and G. E. Zil'berman, Fiz. Tverd. Tela, 1:1697 (1959).
69. A. M. Ratner and G. E. Zil'berman, Fiz. Tverd. Tela, 3:687 (1961).
70. A. M. Ratner, I. A. Rom-Krichevskaya, and Yu. A. Tiunov, in: Quantum Electronics, 1, Naukova Dumka, Kiev (1966), p. 137.
71. A. M. Ratner and V. S. Chernov, in: Quantum Electronics, 2, Naukova Dumka, Kiev (1967), p. 67.
72. A. M. Ratner and V. S. Chernov, Zh. Tekh. Fiz., 38:77 (1968).
73. I. A. Rom-Krichevskaya and A. M. Ratner, Opt. i Spektr., 22:653 (1967).
74. I. A. Rom-Krichevskaya, A. M. Ratner, and A. V. Meshcheryakov, Opt. i Spektr., 19:264 (1965).
75. A. K. Sokolov, and T. N. Zubarev, Fiz. Tverd. Tela, 6:2590 (1964).
76. A. F. Suchkov, Zh. Éksper. i Teoret. Fiz., 49:1495 (1965).
77. K. K. Shvarts, Trudy Instituta Fiz. i Astr. Akad. Nauk Est. SSR, 7:153 (1958).
78. V. N. Tsikunov, Abstract of Candidate's Dissertation, IONKh AN SSSR (1966).
79. G. D. Boyd and J. P. Gordon, Bell System Techn. J., 40:489 (1961).
80. G. D. Boyd and H. Kogelnik, Bell System Techn. J., 41:1347 (1962).
81. G. Bret and F. Gires, Appl. Phys. Lett., 4:75 (1964).
82. D. L. Dexter, J. Chem. Phys., 21:836 (1953); 22(1):1063 (1954).
83. M. M. Ferro, Zs. f. Phys., 56:534 (1929); 57:145 (1929).
84. T. H. Förster Ann. der Phys., 2:55 (1948).
85. A. G. Fox and T. Li, Proc. IRE, 48:1904 (1960).
86. A. G. Fox and T. Li, Bell System Techn. J., 40:453 (1961).
87. G. Gehrer and D. Röss, Zs. f. Naturf., 202:701 (1965).
88. H. Haken and H. Sauermann, Zs. f. Phys., 173:261 (1963).
89. R. Hilsch, Zs. f. Phys., 44:860 (1927).
90. A. Jariv and J. P. Gordon, Proc. IEEE, 51:4 (1963).

91. R. E. Johnson, a.o., Proc. IRE, 49:1942 (1961).
92. W. Kaiser, G. Garret, and D. Wood, Phys. Rev., 123:766 (1961).
93. M. Katzman and J. W. Strozyk, J. Appl. Phys., 35:725 (1964).
94. V. Korennmann, Phys. Rev. Lett., A14:293 (1965).
95. T. Li and S. D. Sims, Proc. IRE, 50:464 (1962).
96. T. Linn and J. Free, Appl. Optics, 4:1099 (1965).
97. T. H. Maiman, Phys. Rev., 123:1145 (1961).
98. F. J. McClung and D. Weiner, IEEE J. Quant. Electr., 1:94 (1965).
99. D. Ross and H. Gehrer, Proc. IEEE, 52:1359 (1964).
100. F. P. Schäfer and W. Schmidt, Zs. f. Naturf., 19a:1019 (1964).
101. A. L. Schawlow and C. H. Townes, Phys. Rev., 112:1940 (1958).
102. E. Snitzer, App. Optics, 5:1487 (1966).
103. P. P. Sorokin, a.o., Phys. Rev., 127:503 (1962).
104. H. Statz, G. A. de Mars, and D. T. Wilson, J. Appl. Phys., 36:1510 (1965).
105. H. Statz and C. L. Tang, Appl. Phys., 36:1816 (1965).
106. C. L. Tang, a.o., J. Appl. Phys., 34:2289 (1963).
107. F. Williams, Phys. Rev., 104:1245 (1956).
108. A. M. Ratner, Lasers Having a Large Angular Divergence, Naukova Dumka, Kiev (1970).
109. B. L. Livshits and A. T. Torsunov, Zh. Éksper. i Teoret. Fiz., 52:1472 (1967).
110. V. V. Korobkin and M. Ya. Shchelev, Zh. Tekh. Fiz., 38:497 (1968).
111. S. A. Mikhnov, Zh. Priklad. Spektr., 7:671 (1967).
112. I. M. Korzhenevich and A. M. Ratner, in: Quantum Electronics, Vol. 4 [in Russian], Naukova Dumka, Kiev (1969), p. 151; Ukrainskii Fiz. Zh., 15, No. 2 (1970).
113. A. M. Ratner, Ukrainskii Fiz. Zh., 13: 1139 (1968).
114. A. M. Ratner, V. S. Solov'ev, and T. I. Tiunova, Zh. Éksper. i Teoret. Fiz., 55:64 (1968).
115. V. D. Kotsubanov, Yu. V. Naboikin, A. M. Ratner, and I. A. Rom-Krichevskaya, Zh. Éksper. i Teoret. Fiz., 53:809 (1967).
116. É. G. Berzing, Yu. V. Naboikin, I. A. Rom-Krichevskaya, and Yu. A. Tiunov, Opt. i Spektr., 22:503 (1967).
117. J. D. Axe and P. P. Sorokin, Phys. Rev., 130:945 (1963).
118. W. V. Smith and P. P. Sorokin, The Laser, McGraw-Hill, New York (1966).
119. J. C. Slater, Quantum Theory of Matter, McGraw-Hill, New York (1951).
120. A. Kiel, in: Quantum Electronics III, (P. Grivet and N. Bloembergen, eds.,) Columbia University Press, New York (1961).
121. G. H. Dieke, in: Advances in Quantum Electronics (J. R. Singer, ed.), Columbia University Press, New York (1961).
122. A. L. Schawlow, in: Advances in Quantum Electronics, (J. R. Singer, ed.), Columbia University Press, New York (1961).
123. A. L. Schawlow, in: Quantum Electronics III (P. Grivet and N. Bloembergen, ed.), Columbia University Press (1961).
124. A. E. Siegman, Lasers and Masers McGraw-Hill, New York (1968).
125. C. M. Stickley, Appl. Optics, 3:967 (1964).
126. E. Gregor, Appl. Optics, 7:2138 (1968).
127. G. D. Currie, Appl. Optics, 8:1069 (1969); Rev. Sci. Instruments, 40:1342 (1969).

128. J. Freud, Appl. Phys. Letters, 12:388 (1968).
129. R. C. Greenhow and A. J. Schmidt, Appl. Phys. Letters, 12:390 (1968).
130. D. R. Dean, L. O. Braun, and R. J. Collins, Appl. Phys. Letters, 12:392 (1968).
131. W. W. Rigrod, Appl. Phys. Letters, 2:51 (1963).
132. H. Statz and G. de Mars, in: Quantum Electronics, Columbia University Press, New York (1960).
133. G. Makhov, J. Appl. Phys., 33:202 (1962).
134. D. M. Sinnett, J. Appl. Phys., 33:1578 (1962).
135. E. Bernal, J. F. Ready, and D. Chen, Proc. IEEE, 52:710 (1964).
136. R. W. Hellworth, Phys. Rev. Letters, 6:39 (1961).
137. C. H. Townes, in: Advances in Quantum Electronics (J. R. Singer, ed.), Columbia University Press, New York (1961).
138. F. J. McClung and R. W. Hellworth, J. Appl. Phys., 33:828 (1962).
139. W. G. Wagner and B. A. Lengyel, J. Appl. Phys., 34:2040 (1963).
140. M. Menat, J. Appl. Phys. 36:73 (1965).
141. C. C. Wang, Proc. IEEE 51:1767 (1963).
142. B. A. Lengyel, Introduction to Laser Physics, Wiley, New York (1966).
143. P. P. Sorokin, J. J. Luzzi, J. R. Lankard, and G. D. Pettit, IBM J. Res. Development, 8:182 (1964).
144. B. H. Soffer, J. Appl. Phys., 35:2551 (1964).
145. M. Hercher, Appl. Optics, 9:947 (1967).
146. M. Di Domenico, J. Appl. Phys., 35:2870 (1964).
147. A. Yariv, Appl. Phys., 36:388 (1965).
148. M. H. Crowell, IEEE Journ. Quantum Elect., QE-1:12 (1965).
149. A. J. DeMaria, D. A. Stetser, and H. Heynan, Appl. Phys. Letters, 8:174 (1966).
150. M. E. Mack, IEEE Journ. Quantum Elect., QE-4:1016 (1968).
151. R. Cubeddu, R. Polloni, C. A. Sacchi and O. Svelto, IEEE Journ. Quantum Elect., QE-5:470 (1969).
152. R. E. McClure, Appl. Phys. Letters 7:148 (1965).
153. T. Ucheda and A. Ueki, IEEE Journ. Quantum Elect., QE-3:17 (1967).
154. W. E. Lamb, Jr., Phys. Rev., 134:A 1420 (1964).
155. F. R. Nash, IEEE Journ. Quantum Elect., QE-3: 189 (1967).
156. F. R. Faxvog, M. S. Thesis, University of Minnesota, Minneapolis, Minnesota.
157. F. R. Faxvog and J. A. Carruthers, Journ. Appl. Phys. Vol. 41 (1970).
158. F. R. Faxvog, C. N. Chow, T. Bieber, and J. A. Carruthers, Appl. Phys. Letters (1970).
159. H. L. Stover and H. Steier, Appl. Phys. Letters, 8:91 (1966).
160. F. R. Faxvog and J. A. Carruthers, J. Appl. Phys. Vol. 42 (1971).
161. L. E. Hargrove, R. L. Fork , and M. A. Pollack, Appl. Phys. Letters, 5:4 (1964).
162. H. Boersch, H. Herziger, H. Lindner , and G. Makosch, Phys. Letters 24A:227 (1967).

INDEX